LECTURES ON VECTOR BUNDLES OVER RIEMANN SURFACES

BY

R. C. GUNNING

PRINCETON UNIVERSITY PRESS

AND THE

UNIVERSITY OF TOKYO PRESS

PRINCETON, NEW JERSEY

1967

Published by Princeton University Press,
Princeton, New Jersey;
in the United Kingdom, by Princeton University Press,
Chichester, West Sussex

5 7 9 10 8 6 4

Printed in the United States of America

ISBN 0-691-07998-6 (pbk)

Preface.

These are notes based on a course of lectures given at Princeton University during the academic year 1966-67. The topic is the analytic theory of complex vector bundles over compact Riemann surfaces. During the preceding academic year, I gave an introductory course on compact Riemann surfaces. The notes for that course have appeared in the same Mathematical Notes series under the title "Lectures on Riemann Surfaces"; they are sufficient, but not necessary, background for reading this set of notes. The present course is not really intended as a natural sequel to the preceding course, though. It is not a systematic presentation of the theory of complex vector bundles, taking up the thread of the discussion of compact Riemann surfaces from the previous year; rather it is a set of lectures on some topics which I found interesting and suggestive of further developments. The aim is to introduce students to an area in which possible research topics lurk, and to provide them with some hunting gear.

In a bit more detail, the topics covered in these lectures are as follows. Sections 1 through 4 contain a general discussion of complex analytic vector bundles over compact Riemann surfaces, from the point of view of sheaf theory. In the preceding course, only sheaves of groups were considered, since that is all that is really needed in one complex variable; but I decided to take this opportunity to introduce the students to some broader classes of sheaves, sheaves of modules over sheaves of rings, and in particular, analytic sheaves on complex manifolds. The relevant definitions, and the connections with complex vector bundles and complex line bundles, are given in section 1; the notion of a coherent analytic sheaf is introduced and discussed in some detail as well. Section 2 contains a discussion of the general structure of coherent analytic sheaves over subdomains of the complex line $\mathbb{C}$, and over the complex projective line $\mathbb{P}$. In section 3 these results are extended to coherent analytic sheaves over arbitrary compact Riemann surfaces, by considering a Riemann surface as a branched covering of $\mathbb{P}$, and examining the behavior of sheaves under such covering mappings. The principal results are the representations of arbitrary coherent analytic sheaves in terms of locally free sheaves, and the existence of meromorphic sections of such sheaves. These results are applied in section 4 to prove the Riemann-Roch theorem for complex analytic vector bundles, and to show the analytic reducibility of vector bundles.

Section 5 is devoted to a rather unsatisfactory descriptive classification of complex analytic vector bundles of rank 2 on a compact Riemann surface. Any such vector bundle can be viewed as an extension of one complex analytic line bundle by another, and the possible extensions are quite easily classified; the difficulty lies in determining which line bundles can be subbundles of a given vector bundle. Mumford's notion of stability comes into the discussion quite naturally here; for unstable bundles the classification can be carried through quite easily, while for stable bundles this approach seems not very satisfactory. No attempt was made to treat stability thoroughly or in detail, since I did not intend to go into the discussion of analytic families of complex vector bundles; that would merit a full year's lectures by itself. The classification was only carried far enough to obtain some results needed for the last part of the course.

Sections 6 through 9 contain a discussion of flat vector bundles over compact Riemann surfaces. There was not time enough to get very far, so this is more an introduction to the subject than a complete discussion; actually, the theory has not yet been developed to the point that a complete discussion is possible. The definition of flat vector bundles and a general description of their relation to complex analytic vector bundles are covered in section 6; the main result is of course Weil's theorem, (Theorem 16). Cohomology with coefficients in a flat sheaf is treated in section 7; and the exact sequence relating this to the cohomology with coefficients in the associated analytic sheaf is introduced in section 8. The concluding section 9 is a preliminary treatment of families of flat vector bundles, including some further details on the analytic equivalence relation among such bundles.

The two appendices cover some questions which came up during the lectures, and which led to brief digressions. The formalism of cohomology with coefficients in a locally free analytic sheaf seemed to be rather confusing at times; the first appendix is an attempt to clarify matters. The analytically trivial flat line bundles on a compact Riemann surface can be described quite directly in terms of the period matrices of the abelian differentials on the surface, while the situation is rather more complicated in the case of vector bundles and the general picture is still incomplete; the second appendix gives an indication of why the vector bundle case is necessarily more complicated.

It must be emphasized that these really are preliminary and informal lecture notes, as claimed on the cover; they are not intended as a complete and polished treatment of the material covered, but rather merely as a set of notes on the lectures, for the convenience of students who attended the course or are interested in this area. Were I to give the same course again, not only would I hope to get much further, but also should I make several changes in the presentation; for instance I would perhaps discuss analytic structures on families of flat vector bundles directly in local terms rather than referring everything back to the characteristic representations of the bundles, following somewhat the lines sketched in other lectures, (see Rice University Pamphlets, vol.54, Fall 1968). It seemed to me, though, that it would be better to make these notes available now for whatever use they might have, rather than to wait for some years to polish and complete the discussion.

I should like to express my thanks here to the students who attended the lectures, for their interest and assistance, and to Elizabeth Epstein, for a beautiful job of typing.

R. C. Gunning

Princeton, New Jersey
July, 1967

Contents

§1. Analytic sheaves.

(a) Sheaves provide a very convenient and useful bit of machinery in complex analysis, and will be used unhesitantly throughout these lectures. Those readers not already familiar with sheaves and their most elementary properties are referred to §2 of last year's Lectures on Riemann Surfaces, which contains all that will be presupposed in this section. Only sheaves of abelian groups were treated there; but more general classes of sheaves are also of importance, so we shall begin by considering some of these.

The definition of a sheaf of rings over a topological space parallels that of a sheaf of abelian groups, except of course that each stalk has the structure of a ring, and that both algebraic operations (addition and multiplication) are continuous. All the rings involved here will be assumed to be commutative, and to possess an identity element. There are thus two canonical sections over any open set, the zero section and the identity section. Considering only the additive structure, a sheaf of rings can be viewed also as a sheaf of abelian groups.

Let $\mathcal{R}$ be a sheaf of rings and $\mathcal{S}$ be a sheaf of abelian groups over a topological space M, with respective projections $\rho: \mathcal{R} \longrightarrow M$ and $\pi: \mathcal{S} \longrightarrow M$. Viewing $\mathcal{R}$ merely as a sheaf of abelian groups, the Cartesian product $\mathcal{R} \times \mathcal{S}$ has the structure of a sheaf of abelian groups over $M \times M$, with the projection $\rho \times \pi: \mathcal{R} \times \mathcal{S} \longrightarrow M \times M$. The restriction of this sheaf to the diagonal $M \subset M \times M$ is then a sheaf of abelian groups over M, which will be denoted by $\mathcal{R} \circ \mathcal{S}$.

<u>Definition.</u> The sheaf $\mathcal{S}$ of abelian groups over M is called a <u>sheaf of modules</u> over the sheaf $\mathcal{R}$ of rings (or more briefly, a sheaf of $\mathcal{R}$-modules) if there is given a sheaf homomorphism $\mathcal{R} \circ \mathcal{S} \longrightarrow \mathcal{S}$ such that for each point $p \in M$ the induced mapping on stalks $\mathcal{R}_p \times \mathcal{S}_p \longrightarrow \mathcal{S}_p$ defines on $\mathcal{S}_p$ the structure of an $\mathcal{R}_p$-module.

For any open set $U \subset M$ a section $f \in \Gamma(U, \mathcal{R} \circ \mathcal{S})$ is readily seen to be the restriction to the diagonal $U \subset U \times U$ of a section $(r,s) \in \Gamma(U \times U, \mathcal{R} \times \mathcal{S})$; that is, there are sections $r \in \Gamma(U, \mathcal{R})$ and $s \in \Gamma(U, \mathcal{S})$ such that $f(p) = (r(p), s(p))$ for all $p \in U$. The sections $\Gamma(U, \mathcal{R})$ form a ring, and the sections $\Gamma(U, \mathcal{S})$ form an abelian group; and the homomorphism $\mathcal{R} \circ \mathcal{S} \longrightarrow \mathcal{S}$ exhibiting $\mathcal{S}$ as a sheaf of $\mathcal{R}$-modules leads to a mapping $\Gamma(U, \mathcal{R} \circ \mathcal{S}) \cong \Gamma(U, \mathcal{R}) \times \Gamma(U, \mathcal{S}) \longrightarrow \Gamma(U, \mathcal{S})$ which clearly defines on $\Gamma(U, \mathcal{S}) = \mathcal{S}_U$ the structure of a $\Gamma(U, \mathcal{R}) = \mathcal{R}_U$-module. Thus $\{\mathcal{S}_U\}$ is in the obvious sense a presheaf of modules over the presheaf $\{\mathcal{R}_U\}$ of rings. Conversely, whenever $\{\mathcal{R}_U\}$ is a presheaf of rings and $\{\mathcal{S}_U\}$ is a presheaf of abelian groups, such that $\mathcal{S}_U$ is an $\mathcal{R}_U$-module and restriction mappings are module homomorphisms, the associated sheaf $\mathcal{S}$ is a sheaf of $\mathcal{R}$-modules.

The notions of sheaf homomorphisms, and related concepts, introduced last year for sheaves of abelian groups, extend readily to sheaves of modules. If $\mathcal{S}$ is a sheaf of $\mathcal{R}$-modules, a subset $\mathcal{J} \subset \mathcal{S}$ is called a <u>subsheaf of $\mathcal{R}$-modules</u> if $\mathcal{J}$ is a subsheaf of abelian groups and, for each point $p \in M$, $\mathcal{J}_p$ is an $\mathcal{R}_p$-submodule of $\mathcal{S}_p$. In the obvious fashion, $\mathcal{R}$ is itself a sheaf of $\mathcal{R}$-modules; a subsheaf $\mathcal{S} \subset \mathcal{R}$ of $\mathcal{R}$-modules is also called a sheaf of ideals in $\mathcal{R}$, since for each point $p \in M$ the stalk $\mathcal{S}_p$ is necessarily an ideal in $\mathcal{R}_p$. If $\mathcal{J} \subset \mathcal{S}$ is a subsheaf of $\mathcal{R}$-modules, the quotient sheaf $\mathcal{S}/\mathcal{J}$ is also a sheaf of $\mathcal{R}$-modules. A sheaf mapping $\varphi: \mathcal{S} \longrightarrow \mathcal{J}$ between two sheaves of $\mathcal{R}$-modules is called an $\mathcal{R}$-homomorphism if for each point $p \in M$ the induced mapping $\varphi_p: \mathcal{S}_p \longrightarrow \mathcal{J}_p$ is a homomorphism of $\mathcal{R}_p$-modules. The kernel $\mathcal{K} \subset \mathcal{S}$ of an $\mathcal{R}$-homomorphism $\varphi: \mathcal{S} \longrightarrow \mathcal{J}$, and the image $\varphi(\mathcal{S}) \subset \mathcal{J}$, are subsheaves of $\mathcal{R}$-modules of their respective sheaves; and φ induces an $\mathcal{R}$-isomorphism $\mathcal{S}/\mathcal{K} \cong \varphi(\mathcal{S})$. An exact sequence of sheaves of $\mathcal{R}$-modules is a sequence

$$\cdots \mathcal{S}_{i-1} \xrightarrow{\varphi_{i-1}} \mathcal{S}_i \xrightarrow{\varphi_i} \mathcal{S}_{i+1} \longrightarrow \cdots$$

of sheaves $\mathcal{S}_i$ of $\mathcal{R}$-modules and of $\mathcal{R}$-homomorphisms φ_i such that for each i the image of φ_{i-1} is precisely the kernel of φ_i . A short exact sequence is an exact sequence of the form

$$0 \longrightarrow \mathcal{S}_1 \xrightarrow{\varphi_1} \mathcal{S}_2 \xrightarrow{\varphi_2} \mathcal{S}_3 \longrightarrow 0 ,$$

where 0 denotes the zero sheaf of $\mathcal{R}$-modules; it is an equivalent way of writing the $\mathcal{R}$-isomorphism $\mathcal{S}_3 \cong \mathcal{S}_2/\varphi_1(\mathcal{S}_1)$.

One additional construction of some importance should be mentioned as well. If $\mathcal{S}$ and $\mathcal{J}$ are sheaves of $\mathcal{R}$-modules, their tensor product $\mathcal{S} \otimes_{\mathcal{R}} \mathcal{J}$ is the sheaf defined by the presheaf $\{\mathcal{S}_U \otimes_{\mathcal{R}_U} \mathcal{J}_U\}$; this is also a sheaf of $\mathcal{R}$-modules, (recalling that all rings considered here are commutative). It is a simple matter to verify that for each point $p \in M$ the stalk $(\mathcal{S} \otimes_{\mathcal{R}} \mathcal{T})_p = \mathcal{S}_p \otimes_{\mathcal{R}_p} \mathcal{T}_p$. For notational convenience, the tensor product will be denoted by $\mathcal{S} \otimes \mathcal{T}$ when there is no danger of confusion. Tensor products of sheaves of modules satisfy many of the familiar properties of tensor products of modules, as is readily verified by considering the separate stalks. In particular, for any sheaf $\mathcal{T}$ of $\mathcal{R}$-modules, $\mathcal{R} \otimes_{\mathcal{R}} \mathcal{T} = \mathcal{T}$, (recalling that all rings considered here have units); and if

$$0 \longrightarrow \mathcal{S}_1 \longrightarrow \mathcal{S}_2 \longrightarrow \mathcal{S}_3 \longrightarrow 0$$

is an exact sequence of sheaves of $\mathcal{R}$-modules, then tensoring with $\mathcal{T}$ yields an exact sequence

$$\mathcal{S}_1 \otimes_{\mathcal{R}} \mathcal{T} \longrightarrow \mathcal{S}_2 \otimes_{\mathcal{R}} \mathcal{T} \longrightarrow \mathcal{S}_3 \otimes_{\mathcal{R}} \mathcal{T} \longrightarrow 0 .$$

(Note especially that it is not claimed that the latter sequence is a full short exact sequence; for if $0 \longrightarrow \mathcal{S}_1 \longrightarrow \mathcal{S}_2$ is exact, it is not necessarily true that $0 \longrightarrow \mathcal{T} \otimes \mathcal{S}_1 \longrightarrow \mathcal{T} \otimes \mathcal{S}_2$ is exact. The sheaf $\mathcal{T}$ is called <u>flat</u> if
$0 \longrightarrow \mathcal{T} \otimes \mathcal{S}_1 \longrightarrow \mathcal{T} \otimes \mathcal{S}_2$ is exact whenever
$0 \longrightarrow \mathcal{S}_1 \longrightarrow \mathcal{S}_2$ is exact.)

(b) If $\mathcal{S}$ and $\mathcal{T}$ are sheaves of $\mathcal{R}$-modules, their <u>direct sum</u> is the sheaf of $\mathcal{R}$-modules defined by the presheaf $\{\mathcal{S}_U \oplus \mathcal{T}_U\}$, and will be denoted by $\mathcal{S} \oplus \mathcal{T}$; this sheaf can be identified in the obvious manner with the sheaf $\mathcal{S} \circ \mathcal{T}$ considered earlier. A particularly simple example is the sheaf of $\mathcal{R}$-modules $\mathcal{R}^m = \mathcal{R} \oplus \ldots \oplus \mathcal{R}$, the direct sum of m copies of the sheaf

$\mathcal{R}$; any sheaf of $\mathcal{R}$-modules isomorphic to $\mathcal{R}^m$ will be called a <u>free sheaf</u> of $\mathcal{R}$-modules <u>of rank</u> m .

Note that the stalk $\mathcal{R}_p^m$ at a point p is just the set of m-tuples of elements of $\mathcal{R}_p$; and elements $R \in \mathcal{R}_p^m$ will be considered as column vectors

$$R = \begin{pmatrix} r_1 \\ r_2 \\ \cdots \\ r_m \end{pmatrix}$$

where $r_j \in \mathcal{R}_p$. The sheaf $\mathcal{R}^m$ has m canonical sections $E_j \in \Gamma(M, \mathcal{R}^m)$, where E_j is the column vector with j-th entry the identity section $1 \in \Gamma(M, \mathcal{R})$ and all other entries the zero section $0 \in \Gamma(M, \mathcal{R})$. These sections are free generators of the sheaf $\mathcal{R}^m$, in the sense that any element $R \in \mathcal{R}_p^m$ can be written uniquely in the form $R = r_1E_1(p) + \ldots + r_mE_m(p)$ for some elements $r_j \in \mathcal{R}_p$; in a similar sense, they are free generators for the module $\Gamma(M, \mathcal{R}^m)$ of sections of the sheaf $\mathcal{R}^m$.

An $\mathcal{R}$-homomorphism $\varphi\colon \mathcal{R}^m \longrightarrow \mathcal{R}^n$ can be described very simply as follows. The image φE_j of the section $E_j \in \Gamma(M, \mathcal{R}^m)$ is a section of $\mathcal{R}^n$, hence can be written uniquely as $\varphi E_j = \sum_{i=1}^{n} f_{ij}E_i$ for some sections $f_{ij} \in \Gamma(M, \mathcal{R})$. These sections (f_{ij}) can be viewed as forming a matrix Φ over the ring $\mathcal{R}_M = \Gamma(M, \mathcal{R})$; this matrix will be called the <u>matrix representing the homomorphism</u> φ . The matrix fully determines the homomorphism. For any element $R \in \mathcal{R}_p^m$ can be represented in the form $R = \sum_{j=1}^{m} r_jE_j(p)$ for some $r_j \in \mathcal{R}_p$; and since φ

is an $\mathcal{R}$-homomorphism, $\varphi(R) = \sum_j r_j \cdot \varphi E_j(p) = \sum_{i,j} r_j f_{ij}(p) E_i(p)$. In other words, using matrix terminology, $\varphi(R) = \Phi(p) \cdot R$. Conversely any $n \times m$ matrix Φ over the ring $\mathcal{R}_M$ determines an $\mathcal{R}$-homomorphism $\varphi\colon \mathcal{R}^m \longrightarrow \mathcal{R}^n$. If $\varphi\colon \mathcal{R}^m \longrightarrow \mathcal{R}^n$ and $\psi\colon \mathcal{R}^n \longrightarrow \mathcal{R}^q$ are two $\mathcal{R}$-homomorphisms, represented by matrices Φ and Ψ respectively, then it is evident that the composition $\psi \circ \varphi$ is represented by the matrix product $\Psi\Phi$. Consequently the homomorphism $\varphi\colon \mathcal{R}^m \longrightarrow \mathcal{R}^m$ is an isomorphism if and only if the matrix Φ representing that homomorphism is an invertible matrix over the ring $\mathcal{R}_M$. (The matrix Φ is of course invertible precisely when its determinant is a unit of the ring $\mathcal{R}_M$, that is, is invertible in the ring $\mathcal{R}_M$. The set of all $m \times m$ invertible matrices over the ring $\mathcal{R}_M$ form a group which will be denoted by $GL(m, \mathcal{R}_M)$, and called the <u>general linear group</u> over the ring $\mathcal{R}_M$.)

A sheaf $\mathcal{S}$ of $\mathcal{R}$-modules over the topological space M is called a <u>locally free sheaf</u> of $\mathcal{R}$-modules <u>of rank</u> m if for some open covering $\mathcal{U} = \{U_\alpha\}$ of M the restrictions $\mathcal{S}|U_\alpha$ are free sheaves of rank m over the various sets U_α. The above description of homomorphisms of free sheaves can be used to derive a convenient description of locally free sheaves; for this purpose the following bit of machinery is required.

If $\mathcal{R}$ is a sheaf of rings over a topological space M, let $\mathcal{GL}(m, \mathcal{R})$ be the sheaf of groups defined by the presheaf $\{GL(m, \mathcal{R}_U)\}$; this is a sheaf of not-necessarily abelian groups of course, but there is no difficulty in the way of the definition, and since nothing is needed but the notation, no further fuss will be made. Let $\mathcal{U} = \{U_\alpha\}$ be an open covering of the space M. By a one-cocycle of $\mathcal{U}$ with coefficients in $\mathcal{GL}(m, \mathcal{R})$ is meant a collection of elements $\Phi_{\alpha\beta} \in GL(m, \mathcal{R}_{U_\alpha \cap U_\beta})$ such that for any triple intersection $U_\alpha \cap U_\beta \cap U_\gamma$, $(\rho\Phi_{\alpha\beta})(\rho\Phi_{\beta\gamma}) = (\rho\Phi_{\alpha\gamma})$, where ρ denotes the natural restriction homomorphism into the group $GL(m, \mathcal{R}_{U_\alpha \cap U_\beta \cap U_\gamma})$. The set of all these one-cocycles will be denoted by $Z^1(\mathcal{U}, \mathcal{GL}(m, \mathcal{R}))$. Two one-cocyles $(\Phi_{\alpha\beta})$, $(\Psi_{\alpha\beta})$ in $Z^1(\mathcal{U}, \mathcal{GL}(m, \mathcal{R}))$ will be called equivalent if there are elements $\Theta_\alpha \in GL(m, \mathcal{R}_{U_\alpha})$ such that $\Psi_{\alpha\beta} = (\rho\Theta_\alpha)\Phi_{\alpha\beta}(\rho\Theta_\beta)^{-1}$, where ρ now denotes the natural restriction homomorphism into the group $GL(m, \mathcal{R}_{U_\alpha \cap U_\beta})$. It is a straightforward matter to show that this is an equivalence relation; the set of equivalence classes will be denoted by $H^1(\mathcal{U}, \mathcal{GL}(m, \mathcal{R}))$, and will be called the first cohomology set of $\mathcal{U}$ with coefficients in $\mathcal{GL}(m, \mathcal{R})$. Suppose that $\mathcal{V} = \{V_\alpha\}$ is another open covering of M, which is a refinement of $\mathcal{U}$ with refining mapping μ; that is, suppose that $\mu: \mathcal{V} \longrightarrow \mathcal{U}$ is a mapping such that $V_\alpha \subset \mu V_\alpha$ for each $V_\alpha \in \mathcal{V}$. Then μ induces a mapping

$$\mu: Z^1(\mathcal{U}, \mathcal{GL}(m, \mathcal{R})) \longrightarrow Z^1(\mathcal{V}, \mathcal{GL}(m, \mathcal{R})),$$

the mapping which takes a cocycle $\Phi_{U_\alpha U_\beta} \in Z^1(\mathcal{U}, \mathcal{GL}(m, \mathcal{R}))$ into the cocyle

$$(\mu\Phi)_{V_\alpha V_\beta} = \rho_{V_\alpha \cap V_\beta} \Phi_{\mu V_\alpha, \mu V_\beta},$$

where $\rho_{V_\alpha \cap V_\beta}$ denotes the restriction mapping to $V_\alpha \cap V_\beta \subset \mu V_\alpha \cap \mu V_\beta$. It is clear that the image is again a cocyle and that μ takes equivalent cocycles into equivalent cocycles;

so that μ induces a mapping

$$\mu^*: H^1(\mathfrak{U}, \mathcal{GL}(m, \mathcal{R})) \longrightarrow H^1(\mathfrak{V}, \mathcal{GL}(m, \mathcal{R})) .$$

Lemma 1. If $\mathfrak{V}$ is a refinement of $\mathfrak{U}$, and if μ and ν are two refining mappings, then $\mu^* = \nu^*$.

Proof. Considering any cocycle $\Phi_{U_\alpha U_\beta} \in Z^1(\mathfrak{U}, \mathcal{GL}(m, \mathcal{R}))$, for each set $V_\alpha \in \mathfrak{V}$ define $\Theta_{V_\alpha} = \rho_{V_\alpha} \Phi_{\mu V_\alpha, \nu V_\alpha}$; this is a well-defined element of $GL(m, \mathcal{R}_{V_\alpha})$, since $V_\alpha \subset \mu V_\alpha \cap \nu V_\alpha$. Then $(\mu\Phi)_{V_\alpha V_\beta} = \rho_{V_\alpha \cap V_\beta} \Phi_{\mu V_\alpha, \mu V_\beta} = \rho_{V_\alpha \cap V_\beta}(\Phi_{\mu V_\alpha, \nu V_\alpha} \Phi_{\nu V_\alpha, \nu V_\beta} \Phi_{\nu V_\beta, \mu V_\beta}) = \rho_{V_\alpha \cap V_\beta}(\Theta_{V_\alpha} \Phi_{\nu V_\alpha, \nu V_\beta} \Theta_{V_\beta}^{-1}) = (\rho\Theta_{V_\alpha})(\nu\Phi)_{V_\alpha V_\beta}(\rho\Theta_{V_\beta})^{-1}$, so that $\mu\Phi$ and $\nu\Phi$ are equivalent; hence $\mu^* = \nu^*$, as asserted.

Now for any two coverings $\mathfrak{U}, \mathfrak{V}$ of M, write $\mathfrak{V} < \mathfrak{U}$ if $\mathfrak{V}$ is a refinement of $\mathfrak{U}$; the set of all coverings is partially ordered under this relation, and by Lemma 1 there is a well-defined mapping $H^1(\mathfrak{U}, \mathcal{GL}(m, \mathcal{R})) \longrightarrow H^1(\mathfrak{V}, \mathcal{GL}(m, \mathcal{R}))$ whenever $\mathfrak{V} < \mathfrak{U}$. It is clear that these mappings are transitive, so that it is possible to introduce the direct limit set

$$H^1(M, \mathcal{GL}(m, \mathcal{R})) = \text{dir.lim.}_{\mathfrak{U}} H^1(\mathfrak{U}, \mathcal{GL}(m, \mathcal{R})),$$

which is called the first cohomology set of M with coefficients in $\mathcal{GL}(m, \mathcal{R})$. (Recall that to define this direct limit introduce the union $\cup_{\mathfrak{U}} H^1(\mathfrak{U}, \mathcal{GL}(m, \mathcal{R}))$; two elements $f \in H^1(\mathfrak{U}, \mathcal{GL}(m, \mathcal{R}))$ and $g \in H^1(\mathfrak{V}, \mathcal{GL}(m, \mathcal{R}))$ will be called equivalent if there is a common refinement $\mathfrak{W} < \mathfrak{U}$, $\mathfrak{W} < \mathfrak{V}$ in which f and g have the same image, and the set of equivalence classes is the direct limit.)

With this machinery at hand, returning to the original question, consider a locally free sheaf $\mathcal{S}$ of $\mathcal{R}$-modules, of rank m ; and let $\mathcal{U} = \{U_\alpha\}$ be an open covering of M such that $\mathcal{S}|U_\alpha$ is free for each U_α. Thus for each set U_α there is an $\mathcal{R}$-isomorphism $\varphi_\alpha: \mathcal{S}|U_\alpha \longrightarrow \mathcal{R}^m|U_\alpha$; and restricting to an intersection $U_\alpha \cap U_\beta \neq \emptyset$, it follows that there is an isomorphism $\varphi_{\alpha\beta} = \varphi_\alpha\varphi_\beta^{-1}: \mathcal{R}^m|U_\alpha \cap U_\beta \longrightarrow \mathcal{R}^m|U_\alpha \cap U_\beta$, and these isomorphisms obviously satisfy the identity $\varphi_{\alpha\beta}\varphi_{\beta\gamma} = \varphi_{\alpha\gamma}$ when restricted to $U_\alpha \cap U_\beta \cap U_\gamma$. Letting $\Phi_{\alpha\beta} \in GL(m, \mathcal{R}_{U_\alpha \cap U_\beta})$ be the matrices representing these isomorphisms of free sheaves, it is thus evident that these matrices form a one-cocycle $(\Phi_{\alpha\beta}) \in Z^1(\mathcal{U}, \mathcal{GL}(m, \mathcal{R}))$. The isomorphisms φ_α are of course not unique, the most general such being of the form $\theta_\alpha\varphi_\alpha$ where $\theta_\alpha: \mathcal{R}^m|U_\alpha \longrightarrow \mathcal{R}^m|U_\alpha$ is an arbitrary isomorphism. Letting $\Theta_\alpha \in GL(m, \mathcal{R}_{U_\alpha})$ be the matrix representing the isomorphism θ_α, the most general one-cocycle associated to the sheaf $\mathcal{S}$ is then of the form $(\rho\Theta_\alpha)\Phi_{\alpha\beta}(\rho\Theta_\beta)^{-1}$ for arbitrary such Θ_α. All these cocycles are equivalent however, so the construction associates to each locally free sheaf and suitably fine covering $\mathcal{U}$ a unique element $\Phi_{\mathcal{U}}(\mathcal{S}) \in H^1(\mathcal{U}, \mathcal{GL}(m, \mathcal{R}))$; this element is called the <u>coordinate bundle representing the locally free sheaf</u> $\mathcal{S}$. Note that whenever $\mathcal{V}$ is a refinement of $\mathcal{U}$ with a refining mapping $\mu: \mathcal{V} \longrightarrow \mathcal{U}$, then $\Phi_{\mathcal{V}}(\mathcal{S}) = \mu^*\Phi_{\mathcal{U}}(\mathcal{S})$.

<u>Lemma 2.</u> Let $\mathcal{S}$ and $\mathcal{S}'$ be locally free sheaves of rank m over a topological space M, and $\Phi_{\mathcal{U}}(\mathcal{S}) \in H^1(\mathcal{U}, \mathcal{GL}(m, \mathcal{R}))$ and $\Phi_{\mathcal{U}'}(\mathcal{S}') \in H^1(\mathcal{U}', \mathcal{GL}(m, \mathcal{R}))$ be coordinate bundles

representing these sheaves in terms of coverings $\mathcal{U}$ and $\mathcal{U}'$ respectively. Then $\mathcal{S}$ and $\mathcal{S}'$ are isomorphic if and only if there is a common refinement $\mathcal{V}$ of $\mathcal{U}$ and $\mathcal{U}'$ such that $\Phi_{\mathcal{V}}(\mathcal{S}) = \Phi_{\mathcal{V}}(\mathcal{S}')$.

Proof. Let $\mathcal{V}$ be a common refinement of the coverings $\mathcal{U}$ and $\mathcal{U}'$, and let $\varphi_\alpha: \mathcal{S}|V_\alpha \longrightarrow \mathcal{R}^m|V_\alpha$ and $\varphi'_\alpha: \mathcal{S}'|V_\alpha \longrightarrow \mathcal{R}^m|V_\alpha$ be isomorphisms. If there is an isomorphism $\theta: \mathcal{S} \longrightarrow \mathcal{S}'$, then for each set V_α the restriction of θ to V_α leads to an isomorphism

$$\theta_\alpha = \varphi'_\alpha \circ \theta \circ \varphi_\alpha^{-1}: \mathcal{R}^m|V_\alpha \longrightarrow \mathcal{R}^m|V_\alpha .$$

Letting $\Theta_\alpha \in GL(m, \mathcal{R}_{V_\alpha})$ be the matrix representing θ_α, it follows readily that $\Phi'_{\alpha\beta} = (\rho\Theta_\alpha)\Phi_{\alpha\beta}(\rho\Theta_\beta)^{-1}$; thus $(\Phi'_{\alpha\beta})$ and $(\Phi_{\alpha\beta})$ are equivalent, and hence $\Phi_{\mathcal{V}}(\mathcal{S}') = (\Phi'_{\alpha\beta}) = (\Phi_{\alpha\beta}) = \Phi_{\mathcal{V}}(\mathcal{S})$ in $H^1(\mathcal{V}, \mathcal{GL}(m, \mathcal{R}))$. Conversely, if $\Phi_{\mathcal{V}}(\mathcal{S}') = \Phi_{\mathcal{V}}(\mathcal{S})$ there are matrices Θ_α satisfying the above equality, and defining isomorphisms θ_α; and then it follows readily that $\theta = (\varphi'_\alpha)^{-1} \circ \theta_\alpha \circ \varphi_\alpha$ is the desired isomorphism from $\mathcal{S}$ to $\mathcal{S}'$. This concludes the proof.

For any locally free sheaf $\mathcal{S}$ over M, the various coordinate bundles representing $\mathcal{S}$ in all suitably fine coverings all lead to the same element $\Phi(\mathcal{S}) \in H^1(M, \mathcal{GL}(m, \mathcal{R}))$, since $\Phi_{\mathcal{V}}(\mathcal{S}) = \mu^*\Phi_{\mathcal{U}}(\mathcal{S})$ for any refining mapping $\mu: \mathcal{V} \longrightarrow \mathcal{U}$. This element is called the <u>fibre bundle representing the locally free sheaf</u> $\mathcal{S}$.

Theorem 1. If $\mathcal{R}$ is a sheaf of rings over a topological space M, the mapping which associates to a locally free sheaf of $\mathcal{R}$-modules of rank m its representative fibre bundle establishes a one-to-one correspondence between the set of such sheaves and the cohomology set $H^1(M, \mathcal{GL}(m, \mathcal{R}))$.

Proof. It is an immediate consequence of Lemma 2 and the definition of the direct limit that the mapping $\mathcal{S} \longrightarrow \Phi(\mathcal{S})$ is a one-to-one mapping from the set of locally free sheaves of rank m into the cohomology set $H^1(M, \mathcal{GL}(m, \mathcal{R}))$. To complete the proof it is only necessary to show that the mapping is onto. For any given cohomology class $\Phi \in H^1(M, \mathcal{GL}(m, \mathcal{R}))$ select a covering $\mathcal{U}$ and a representative cocycle $(\Phi_{\alpha\beta}) \in Z^1(\mathcal{U}, \mathcal{GL}(m, \mathcal{R}$ for Φ. There is no loss of generality in assuming that $\mathcal{U}$ is a basis for the topology of M, since there is always a basis refining any open covering. Introduce a presheaf $\{\mathcal{S}_{U_\alpha}\}$, $U_\alpha \in \mathcal{U}$, defined as follows: for each U_α put $\mathcal{S}_{U_\alpha} = \mathcal{R}^m | U_\alpha$, and for each containment $U_\beta \subset U_\alpha$ let $\rho_{U_\beta U_\alpha}: \mathcal{S}_{U_\alpha} \longrightarrow \mathcal{S}_{U_\beta}$ be the homomorphism of free sheaves associated to the matrix $\Phi_{\beta\alpha} \in GL(m, \mathcal{R}_{U_\alpha \cap U_\beta})$. The cocycle conditions on the matrices $\Phi_{\alpha\beta}$ show that this is indeed a well-defined presheaf. Letting $\mathcal{S}$ be the sheaf associated to this presheaf, it is immediately evident that $(\Phi_{\alpha\beta})$ represents the coordinate bundle of the sheaf $\mathcal{S}$, hence that $\Phi = \Phi(\mathcal{S})$. This therefore serves to conclude the proof.

As a notational convention, for any locally free sheaf $\mathcal{S}$ of $\mathcal{R}$-modules of rank m over the topological space M, let

$\Phi(\mathcal{S}) \in H^1(M, \mathcal{GL}(m, \mathcal{R}))$ denote the fibre bundle representing that sheaf; and for any sufficiently fine open covering $\mathfrak{U}$ of M, let $\Phi_{\mathfrak{U}}(\mathcal{S}) \in H^1(\mathfrak{U}, \mathcal{GL}(m, \mathcal{R}))$ denote the coordinate bundle representing that sheaf, or equivalently, representing the fibre bundle $\Phi(\mathcal{S})$. Conversely, for any fibre bundle $\Phi \in H^1(M, \mathcal{GL}(m, \mathcal{R}))$, let $\mathcal{R}(\Phi)$ denote the locally free sheaf of $\mathcal{R}$-modules represented by the fibre bundle Φ. The special case $m = 1$ merits a bit further attention. Note that $\mathcal{GL}(1, \mathcal{R})$ is the subsheaf of $\mathcal{R}$ formed of the invertible elements in $\mathcal{R}$, hence is a sheaf of abelian groups; the notation $\mathcal{GL}(1, \mathcal{R}) = \mathcal{R}^* \subset \mathcal{R}$ will sometimes be used as well. In this case, since $\mathcal{R}^*$ is commutative, the cohomology set $H^1(M, \mathcal{R}^*) = H^1(M, \mathcal{GL}(1, \mathcal{R}))$ is a group as well; indeed, this is just the first cohomology group of M with coefficients in the sheaf $\mathcal{R}^*$ of abelian groups, as defined in §3 of last year's Lectures on Riemann Surfaces. The elements of $H^1(M, \mathcal{R}^*)$ will usually be denoted by lower case Greek letters.

If $\mathcal{S}$ and $\mathcal{T}$ are locally free sheaves of $\mathcal{R}$-modules of ranks m and n respectively, it is clear that the direct sum $\mathcal{S} \oplus \mathcal{T}$ is a locally free sheaf of $\mathcal{R}$-modules of rank $m+n$, and that the tensor product $\mathcal{S} \otimes \mathcal{T}$ is a locally free sheaf of $\mathcal{R}$-modules of rank mn. Both operations are associative, and the sheaf $\mathcal{R}$ is the identity for the tensor product operation. Similar operations can be introduced in the cohomology set $H^1(M, \mathcal{GL}(*, \mathcal{R})) = \bigcup_{m=1}^{\infty} H^1(M, \mathcal{GL}(m, \mathcal{R}))$; if $\Phi \in H^1(M, \mathcal{GL}(m, \mathcal{R}))$ and $\Psi \in H^1(M, \mathcal{GL}(n, \mathcal{R}))$, selecting representative cocycles $(\Phi_{\alpha\beta})$ and $(\Psi_{\alpha\beta})$ for a common open covering $\mathcal{U}$ of M, let $\Phi \oplus \Psi$ be the cohomology class defined by the cocycle $(\Phi_{\alpha\beta} \oplus \Psi_{\alpha\beta})$ and $\Phi \otimes \Psi$ be the cohomology class defined by the cocycle $(\Phi_{\alpha\beta} \otimes \Psi_{\alpha\beta})$. The direct sum and tensor product (or Kronecker product) of matrices are defined as usual, and it is easy to see that the definitions are independent of the choices of cocycles. It is clear from the definitions that, considering the fibre bundles representing locally free sheaves, $\Phi(\mathcal{S} \oplus \mathcal{J}) = \Phi(\mathcal{S}) \oplus \Phi(\mathcal{J})$ and $\Phi(\mathcal{S} \otimes \mathcal{J}) = \Phi(\mathcal{S}) \otimes \Phi(\mathcal{J})$. It is left to the reader also to verify that the tensor product is a group operation in the subset of locally free sheaves of rank 1; this corresponds to the group structure in $H^1(M, \mathcal{R}^*) = H^1(M, \mathcal{GL}(1, \mathcal{R}))$. (For this reason, locally free sheaves of rank 1 are also sometimes called invertible sheaves.) Note further that $\mathcal{S} \otimes (\mathcal{J}_1 \oplus \mathcal{J}_2) = (\mathcal{S} \otimes \mathcal{J}_1) \oplus (\mathcal{S} \otimes \mathcal{J}_2)$, as one verifies by considering separate stalks.

It should be remarked in passing that a locally free sheaf is flat; that is to say, if $\mathcal{S}$ is a locally free sheaf of $\mathcal{R}$-modules and $0 \longrightarrow \mathcal{J}_1 \longrightarrow \mathcal{J}_2$ is any exact sequence of sheaves of $\mathcal{R}$-modules, then $0 \longrightarrow \mathcal{S} \otimes \mathcal{J}_1 \longrightarrow \mathcal{S} \otimes \mathcal{J}_2$ is also an exact sequence. (The result is of a local nature, so it suffices to prove it for free sheaves; but since $\mathcal{R}^m \otimes \mathcal{J} = (\mathcal{R} \oplus \ldots \oplus \mathcal{R}) \otimes \mathcal{J} = (\mathcal{R} \otimes \mathcal{J}) \oplus \ldots \oplus (\mathcal{R} \otimes \mathcal{J}) = \mathcal{J} \oplus \ldots \oplus \mathcal{J}$, the assertion is trivially true.)

(c) The principal interest of the preceding constructions for the present lectures lies in their usefulness in studying complex analytic manifolds, and in particular, one-dimensional complex analytic manifolds, (Riemann surfaces). On a Riemann surface M, note that the sheaf $\mathcal{O}$ of germs of holomorphic functions is actually a sheaf of rings, (commutative rings with identity). A sheaf of modules over the sheaf of rings $\mathcal{O}$ is called an <u>analytic sheaf</u> over the Riemann surface.

According to Theorem 1, there is a one-to-one correspondence between the locally free analytic sheaves of rank m over the Riemann surface M and the cohomology set $H^1(M, \mathcal{GL}(m, \mathcal{O}))$. In this case, $\mathcal{GL}(m, \mathcal{O})$ is the sheaf defined by the pre-sheaf $\{GL(m, \mathcal{O}_U)\}$, where $\mathcal{O}_U$ is the ring of holomorphic functions over the open set $U \subset M$; but since $GL(m, \mathcal{O}_U)$ can be viewed as the group of complex analytic mappings from U into the complex analytic manifold $GL(m, \mathbb{C})$, the sheaf $\mathcal{GL}(m, \mathcal{O})$ can be considered as the sheaf of germs of complex analytic mappings from M into $GL(m, \mathbb{C})$. The elements of $H^1(M, \mathcal{GL}(m, \mathcal{O}))$ will be called <u>complex analytic vector bundles</u> of rank m over the Riemann surface M; the special case $m = 1$ was considered last year, and the elements of the set (actually the group) $H^1(M, \mathcal{GL}(1, \mathcal{O})) = H^1(M, \mathcal{O}^*)$ were called <u>complex line bundles.</u> Recall from last year's lectures (in particular §4, page 54), that to any complex line bundle $\xi \in H^1(M, \mathcal{O}^*)$ there was canonically associated the sheaf $\mathcal{O}(\xi)$ of germs of holomorphic cross-sections of that bundle; and the construction given there was precisely the construction used in the proof of

Theorem 1 in the paragraph above. Thus the sheaves $\mathcal{O}(\xi)$ are precisely the locally free analytic sheaves of rank 1 over the Riemann surface M ; and the fibre bundle representing the sheaf $\mathcal{O}(\xi)$ is precisely the complex line bundle ξ . The locally free analytic sheaves of higher ranks thus form a straightforward generalization of the sheaves considered last year, with a parallel notation; and for this reason, a locally free analytic sheaf of rank m represented by a complex analytic vector bundle $\Phi \in H^1(M, \mathcal{GL}(m, \mathcal{O}))$ will sometimes be called the sheaf of germs of holomorphic cross-sections of the vector bundle Φ .

The sheaf $\mathcal{C}$ of continuous complex-valued functions on the Riemann surface M is also a sheaf of rings, as is the sheaf $\mathcal{C}^\infty$ of infinitely differentiable functions; recall the natural inclusions $\mathcal{O} \subset \mathcal{C}^\infty \subset \mathcal{C}$. Locally free sheaves of rank m over the sheaf of rings $\mathcal{C}$ are in one-to-one correspondence with the set $H^1(M, \mathcal{GL}(m, \mathcal{C}))$, called the set of continuous complex vector bundles over M ; and the locally free sheaves of $\mathcal{C}^\infty$-modules of rank m are in one-to-one correspondence with the set $H^1(M, \mathcal{GL}(m, \mathcal{C}^\infty))$, called the set of C^∞ complex vector bundles. The inclusion mappings $\mathcal{O} \longrightarrow \mathcal{C}^\infty \longrightarrow \mathcal{C}$ lead to inclusion mappings $\mathcal{GL}(m, \mathcal{O}) \longrightarrow \mathcal{GL}(m, \mathcal{C}^\infty) \longrightarrow \mathcal{GL}(m, \mathcal{C})$, and these in turn to inclusion mappings

$$H^1(M, \mathcal{GL}(m, \mathcal{O})) \longrightarrow H^1(M, \mathcal{GL}(m, \mathcal{C}^\infty)) \longrightarrow H^1(M, \mathcal{GL}(m, \mathcal{C})) .$$

In the case $m = 1$, these inclusions were used last year in classifying complex line bundles, and hence locally free analytic sheaves of rank 1; (recall in particular §7(a) and §8(a)). For the case $m > 1$ the situation is considerably more complicated, with no satisfactory classification theorem yet in sight; indeed, the idea of a classification theorem for the set of all complex analytic vector bundles of the same form as that for complex line bundles, may not be a reasonable one.

(d) The class of locally free sheaves is not closed under completion of short exact sequences; and many of the analytic sheaves naturally arising, more especially in several complex variables, are not locally free. For these reasons it is convenient to introduce a somewhat broader class of analytic sheaves. For the purpose of the present lectures, it will suffice to consider only analytic sheaves (sheaves of $\mathcal{O}$-modules) over a Riemann surface M.

An analytic sheaf $\mathcal{S}$ is called a <u>coherent analytic sheaf</u> over the Riemann surface M if to each point $p \in M$ there is an open neighborhood U such that the restriction of $\mathcal{S}$ to U is the cokernel of an $\mathcal{O}$-homomorphism of free analytic sheaves, that is to say, such that there is an exact sequence of analytic sheaves of the form

$$(1) \qquad \mathcal{O}^{m_1}|U \xrightarrow{\lambda_1} \mathcal{O}^{m}|U \xrightarrow{\lambda} \mathcal{S}|U \longrightarrow 0 .$$

The canonical sections $E_j \in \Gamma(U, \mathcal{O}^m|U)$ generate the sheaf $\mathcal{O}^m|U$ as a sheaf of $\mathcal{O}$-modules, hence their images $\lambda E_j \in \Gamma(U, \mathcal{S}|U)$ generate the sheaf $\mathcal{S}|U$ as a sheaf of $\mathcal{O}$-modules; so in this sense, a coherent analytic sheaf is <u>finitely generated.</u>

Letting $S_j = \lambda E_j \in \Gamma(U, \mathcal{S}|U)$ be the generators of the sheaf $\mathcal{S}$, the homomorphism λ is described by

$\lambda(\sum_j f_j \cdot E_j(p)) = \sum_j f_j \cdot S_j(p)$ for all m-tuples $(f_j) \in \mathcal{O}_p^m$; the kernel of λ is thus the subsheaf of $\mathcal{O}^m|U$ consisting of those m-tuples $(f_j) \in \mathcal{O}_p^m$ for points $p \in U$ such that $\sum_j f_j \cdot S_j(p) = 0$, and is called the sheaf of relations among the generators S_j. The exact sequence (1) shows that the sheaf of relations is also finitely generated. Thus a coherent analytic sheaf can be described as a finitely generated analytic sheaf such that the sheaf of relations among the generators is also finitely generated. It is convenient to derive a few conditions guaranteeing that an analytic sheaf be coherent; for the sake of simplicity, only the case of one complex variable will be considered.

Lemma 3. (Oka's Lemma) The kernel and image of any homomorphism $\lambda: \mathcal{O}^m \longrightarrow \mathcal{O}^n$ of free sheaves are coherent analytic sheaves.

Proof. Selecting any point p in the Riemann surface, it suffices to find an open neighborhood U of p and a sheaf homomorphism $\lambda_1: \mathcal{O}^{m_1}|U \longrightarrow \mathcal{O}^m|U$ such that the sequence

(2) $$\mathcal{O}^{m_1}|U \xrightarrow{\lambda_1} \mathcal{O}^m|U \xrightarrow{\lambda} \mathcal{O}^n|U$$

is exact. For then the image of λ is the cokernel of λ_1, hence is coherent by definition; and the kernel of λ is the image of λ_1, so is coherent (locally at least, which suffices) by the first observation.

As for the proof of the existence of the homomorphism λ_1, consider first the case $n = 1$; then λ is represented by an $m \times 1$ matrix $\Lambda = (\ell_1,\ldots,\ell_m)$ of holomorphic functions. Choose a coordinate mapping z in an open neighborhood U of p such that $z(p) = 0$. The functions $\ell_j(z)$ can then be written $\ell_j(z) = z^{r_j} \cdot u_j(z)$ for analytic functions $u_j(z)$ such that $u_j(0) \neq 0$; and by taking U sufficiently small, it can be assumed that $u_j(z) \neq 0$ for all $z \in U$. By relabeling, suppose that $r_1 = \min(r_1, \ldots, r_m)$. Then if $(f_j) \in \mathcal{O}_p^m$ is an element of the kernel of λ , necessarily

$$0 = \sum_j \ell_j(z) f_j(z) = z^{r_1} \sum_j z^{r_j - r_1} u_j(z) f_j(z) ,$$

so that

$$f_1(z) = - \frac{1}{u_1(z)} \sum_{j=2}^{m} z^{r_j - r_1} u_j(z) f_j(z) .$$

Consequently, it is evident that $\lambda_1 : \mathcal{O}^{m-1}|U \longrightarrow \mathcal{O}^m|U$, defined as the mapping sending the $(m-1)$-tuple $(g_j) \in \mathcal{O}_p^{m-1}$ to the m-tuple $(f_j) \in \mathcal{O}_p^m$ where

$$f_1(z) = - \frac{1}{u_1(z)} \sum_{j=2}^{m} z^{r_j - r_1} u_j(z) g_j(z) ,$$

$$f_j(z) = g_j(z) \quad \text{for} \quad j = 2, \ldots, m ,$$

is the desired mapping.

The remainder of the proof will be by induction on the index n ; the case $n = 1$ having just been treated, suppose that the result holds also for $n - 1$, and consider the mapping $\lambda : \mathcal{O}^m \longrightarrow \mathcal{O}^n = \mathcal{O}^{n-1} \oplus \mathcal{O}^n$. This homomorphism can be split into the direct sum $\lambda = \lambda' \oplus \lambda''$ of homomorphisms $\lambda' : \mathcal{O}^m \longrightarrow \mathcal{O}^{n-1}$ and $\lambda'' : \mathcal{O}^m \longrightarrow \mathcal{O}^1$; and the kernel of λ is the intersection of the kernels of λ' and of λ'' . By the induction hypothesis, for some open neighborhood U of the point p there will be an exact sequence of analytic sheaves

$$\mathcal{O}^{m_1'}|U \xrightarrow{\lambda_1'} \mathcal{O}^{m}|U \xrightarrow{\lambda'} \mathcal{O}^{n-1}|U$$

for some homomorphism λ_1' ; and by the result proved for the case $n = 1$, by reducing the size of U if necessary, there will be an exact sequence of analytic sheaves

$$\mathcal{O}^{m_1}|U \xrightarrow{\lambda_1''} \mathcal{O}^{m_1'}|U \xrightarrow{\lambda''\circ\lambda_1'} \mathcal{O}^{1}|U$$

for some homomorphism λ_1'' . Then the exact sequence

$$\mathcal{O}^{m_1}|U \xrightarrow{\lambda_1'\circ\lambda_1''} \mathcal{O}^{m}|U \xrightarrow{\lambda'\oplus\lambda''} \mathcal{O}^{n}|U$$

which results serves to conclude the proof.

<u>Lemma 4.</u> If $\mathcal{S}$ and $\mathcal{J}$ are coherent analytic subsheaves of $\mathcal{O}^n$, then $\mathcal{S}\cap\mathcal{J}$ is also a coherent analytic subsheaf of $\mathcal{O}^n$.

Proof. Restricting attention to a sufficiently small open neighborhood U of a point p on the Riemann surface, there are sheaf homomorphisms $\sigma\colon \mathcal{O}^s \longrightarrow \mathcal{O}^n$ and $\tau\colon \mathcal{O}^t \longrightarrow \mathcal{O}^n$ such that $\mathcal{S} = \sigma(\mathcal{O}^s)$ and $\mathcal{J} = \tau(\mathcal{O}^t)$. Consider then the homomorphism $\sigma\oplus\tau\colon \mathcal{O}^{s+t} = \mathcal{O}^s \oplus \mathcal{O}^t \longrightarrow \mathcal{O}^n$, and let $\mathcal{K}$ be its kernel; note that an element $(F,G) \in \mathcal{O}^s_q \oplus \mathcal{O}^t_q$ lies in $\mathcal{K}$ if and only if $\sigma(F) = -\tau(G) \in \mathcal{S}_q \cap \mathcal{J}_q$, hence the map $\sigma'\colon \mathcal{K} \longrightarrow \mathcal{O}^n$ defined by $\sigma'(F,G) = \sigma(F)$ has as its image precisely the subsheaf $\mathcal{S}\cap\mathcal{J}\subset\mathcal{O}^n$. Since $\mathcal{K}$ is coherent by Lemma 3, upon restricting U still further if necessary there will be a homomorphism

$\varphi: \mathcal{O}^m \longrightarrow \mathcal{O}^{s+t}$ with image precisely $\mathcal{K}$. But then $\sigma'\circ\varphi: \mathcal{O}^m \longrightarrow \mathcal{O}^n$ has image precisely $\mathcal{S} \cap \mathcal{T}$, so that $\mathcal{S} \cap \mathcal{T}$ is coherent by Lemma 3, thus concluding the proof.

The following properties of coherent analytic sheaves are straightforward consequence of the definition and of Oka's lemma.

Theorem 2. (a) If $\mathcal{T}$ is a coherent analytic sheaf, then an analytic subsheaf $\mathcal{S} \subset \mathcal{T}$ is coherent if and only if it is locally finitely generated.

(b) If, in an exact sequence of analytic sheaves of the form

$$0 \longrightarrow \mathcal{R} \xrightarrow{\varphi} \mathcal{S} \xrightarrow{\psi} \mathcal{T} \longrightarrow 0,$$

any two of the sheaves are coherent, then the third is also coherent.

(c) If $\mathcal{S}$ and $\mathcal{T}$ are coherent analytic sheaves, then the kernel, image, and cokernel of any sheaf homomorphism $\varphi: \mathcal{S} \longrightarrow \mathcal{T}$ are also coherent analytic sheaves.

(d) If $\mathcal{R}$ and $\mathcal{S}$ are coherent analytic subsheaves of a coherent analytic sheaf $\mathcal{T}$, then $\mathcal{R} + \mathcal{S}$ and $\mathcal{R} \cap \mathcal{S}$ are also coherent.

(e) If $\mathcal{R}$ and $\mathcal{S}$ are coherent analytic sheaves, so are $\mathcal{R} \oplus \mathcal{S}$ and $\mathcal{R} \otimes_{\mathcal{O}} \mathcal{S}$.

Proof. (a) It is only necessary to show that an analytic subsheaf $\mathcal{S} \subset \mathcal{T}$ which is locally finitely generated is coherent. Restricting attention to a sufficiently small open neighborhood U of a point p, the hypothesis is that $\mathcal{S}$ is the image of a sheaf homomorphism $\varphi: \mathcal{O}^m \longrightarrow \mathcal{S}$; so clearly it suffices to show that the kernel $\mathcal{K}$ of φ is finitely generated. Since $\mathcal{S}$ is coherent, further restricting U if necessary, there are homomorphisms σ, σ_1 such that the vertical column in the accompanying diagram is an exact sequence.

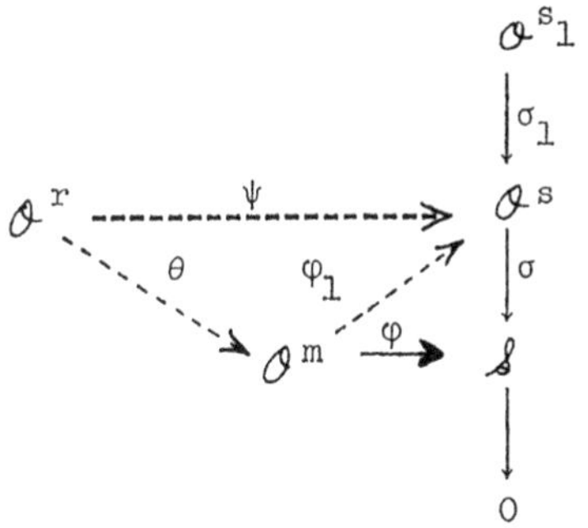

There is a homomorphism φ_1 such that the above diagram is commutative, reducing U further if necessary; (this is left as an easy exercise to the reader - recall that mappings of free sheaves are determined by what they do to the canonical sections E_j). By commutativity, an element $F \in \mathcal{O}^m_q$ is in the kernel $\mathcal{K}$ of φ if and only if $\sigma\varphi_1(F) = 0$, that is, if and only if $\varphi_1(F) \in$ kernel σ = image σ_1 . Since image $\varphi_1 \cap$ image σ_1 is coherent by Lemmas 3 and 4, there is a sheaf homomorphism $\psi: \mathcal{O}^r \longrightarrow \mathcal{O}^s$ such that image ψ = image $\varphi_1 \cap$ image σ_1 ; and since image $\psi \subset$ image φ_1 , there is a sheaf homomorphism $\theta: \mathcal{O}^r \longrightarrow \mathcal{O}^m$ such that the above diagram commutes. Now $F \in \mathcal{K}$ if and only if $\varphi_1(F) = \psi(G)$ for some element $G \in \mathcal{O}^r$, hence if and only if $\varphi_1(F - \theta(G)) = 0$; thus $\mathcal{K}$ = (image θ) + (kernel φ_1) , and $\mathcal{K}$ is finitely generated since image θ and kernel φ_1 are both coherent by Lemma 3.

(b) First suppose that $\mathcal{R}$ and $\mathcal{S}$ are coherent; then over a sufficiently small open neighborhood U there are sheaf homomorphisms $\rho\colon \mathcal{O}^r \longrightarrow \mathcal{R}$ and $\sigma\colon \mathcal{O}^s \longrightarrow \mathcal{S}$, both of which are onto. Restricting U if necessary, there is clearly a homomorphism φ_1 such that the following diagram is commutative and exact.

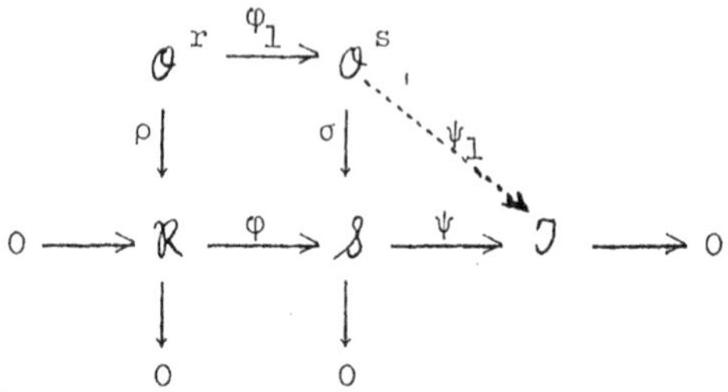

Letting $\psi_1 = \psi\sigma$, it is clear that $\mathcal{T}$ is the image of ψ_1; an easy diagram chase shows that the kernel of ψ_1 is (kernel ψ_1) = = (image φ_1) + (kernel σ), hence (kernel ψ_1) is finitely generated since both (image φ_1) and (kernel σ) are coherent. Thus $\mathcal{T}$ is coherent.

Next suppose that $\mathcal{S}$ and $\mathcal{T}$ are coherent; for a suitably small open neighborhood U there is a sheaf homomorphism $\sigma\colon \mathcal{O}^s \longrightarrow \mathcal{S}$ which is onto.

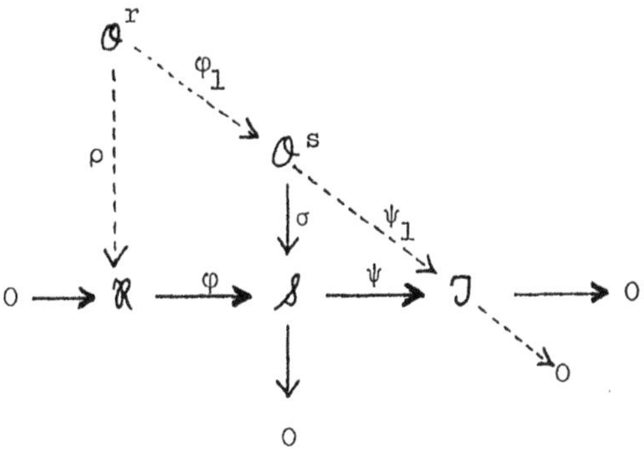

Letting $\psi_1 = \psi\sigma$, the image of ψ_1 is coherent by part (a); indeed, the proof of that part shows that the kernel of ψ_1 is finitely generated, so upon restricting U further if necessary, there will be a sheaf homomorphism $\varphi_1\colon \mathcal{O}^r \longrightarrow \mathcal{O}^s$ making the diagonal sequence in the above diagram exact. Since the image of $\sigma\varphi_1$ lies in the kernel of ψ (or the image of φ), there is a sheaf homomorphism $\rho\colon \mathcal{O}^r \longrightarrow \mathcal{R}$ so that the diagram remains commutative; needless to say, U may have to be reduced still further. Clearly then ρ is onto, so that $\mathcal{R}$ is a finitely generated subsheaf of $\mathcal{S}$, hence $\mathcal{R}$ is coherent by part (a).

Finally suppose that $\mathcal{R}$ and $\mathcal{T}$ are coherent; for a suitably small open neighborhood U there are sheaf homomorphisms $\rho\colon \mathcal{O}^r \longrightarrow \mathcal{R}$ and $\tau\colon \mathcal{O}^t \longrightarrow \mathcal{T}$ which are both onto.

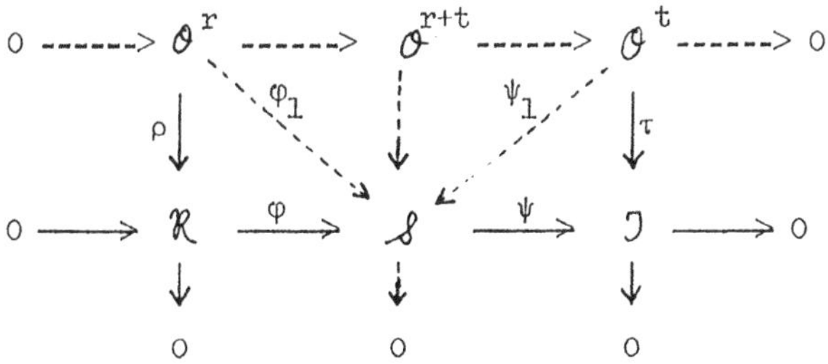

Let $\varphi_1 = \varphi\rho$; and let ψ_1: $\mathcal{O}^t \longrightarrow \mathcal{S}$ be a sheaf homomorphism such that the above diagram is commutative. It is then clear that the mapping $\varphi_1 \oplus \psi_1$: $\mathcal{O}^{r+t} = \mathcal{O}^r \oplus \mathcal{O}^t \longrightarrow \mathcal{S}$ is onto, so the full diagram above is exact and commutative. This shows that $\mathcal{S}$ is finitely generated; and repeating the argument on the kernels in the exact sequence of the top row shows that $\mathcal{S}$ is actually coherent.

(c) Since $\mathcal{S}$ is locally finitely generated, so is the image $\varphi(\mathcal{S}) \subset \mathcal{J}$; but then from part (a) it follows that $\varphi(\mathcal{S})$ is coherent. Considering next the exact sequences

$$0 \longrightarrow (\text{kernel } \varphi) \longrightarrow \mathcal{S} \longrightarrow \varphi(\mathcal{S}) \longrightarrow 0$$

and

$$0 \longrightarrow \varphi(\mathcal{S}) \longrightarrow \mathcal{S} \longrightarrow (\text{cokernel } \varphi) \longrightarrow 0 ,$$

it follows from (b) that both the kernel of φ and the cokernel of φ are coherent.

(d) Note that $\mathcal{R} + \mathcal{S}$ is clearly locally finitely generated, hence is coherent by (a). Note further that $\mathcal{J}/\mathcal{R}$ is coherent by (b); so considering the natural homomorphism φ: $\mathcal{J} \longrightarrow \mathcal{J}/\mathcal{R}$, the image of $\mathcal{S}$ will be a locally finitely generated subsheaf of $\mathcal{J}/\mathcal{R}$, hence is coherent by (a). Since $\varphi(\mathcal{S}) = \mathcal{S}/\mathcal{S} \cap \mathcal{R}$, it follows from (b) that $\mathcal{S} \cap \mathcal{R}$ is coherent also.

(e) Since $0 \longrightarrow \mathcal{R} \longrightarrow \mathcal{R} \oplus \mathcal{S} \longrightarrow \mathcal{S} \longrightarrow 0$, it follows immediately from (b) that $\mathcal{R} \oplus \mathcal{S}$ is coherent. Over some open neighborhood U of any point there will be an exact sequence of analytic sheaves of the form

$$\mathcal{O}^{r_1} \xrightarrow{\rho_1} \mathcal{O}^r \xrightarrow{\rho} \mathcal{R} \longrightarrow 0 .$$

Then tensoring with $\mathcal{S}$ yields the exact sequence

$$\mathcal{O}^{r_1} \otimes_{\mathcal{O}} \mathcal{S} \longrightarrow \mathcal{O}^r \otimes_{\mathcal{O}} \mathcal{S} \longrightarrow \mathcal{R} \otimes_{\mathcal{O}} \mathcal{S} \longrightarrow 0 .$$

Now noting that $\mathcal{O}^{r_1} \otimes_{\mathcal{O}} \mathcal{S} \cong \mathcal{S}^{r_1}$ and $\mathcal{O}^r \otimes_{\mathcal{O}} \mathcal{S} \cong \mathcal{S}^r$, and that $\mathcal{S}^{r_1}$ and $\mathcal{S}^r$ are both coherent, it follows from (c) that $\mathcal{R} \otimes_{\mathcal{O}} \mathcal{S}$ is coherent, as the cokernel of a homomorphism of coherent analytic sheaves. With this observation, the proof of the theorem is concluded.

§2. Local structure of coherent analytic sheaves

(a) Over a Riemann surface, any coherent analytic sheaf can be described quite simply in terms of complex vector bundles; the present section will be devoted to a local and semi-local version of this relationship on a general Riemann surface, and the global version over the complex projective line.

Theorem 3. On a Riemann surface, every coherent analytic subsheaf of a locally free sheaf is locally free.

Proof. The assertion being local, it suffices to prove that a coherent analytic subsheaf $\mathcal{S} \subset \mathcal{O}^m$ is locally free. For any point p on the surface, there is an open neighborhood U of p over which the sheaf is finitely generated; thus there are sections $S_i \in \Gamma(U, \mathcal{O}^m)$, $i = 1,\ldots,r$, such that the sheaf homomorphism $\varphi: \mathcal{O}^r|U \longrightarrow \mathcal{O}^m|U$, defined by $(f_1,\ldots,f_r) \in \mathcal{O}^r_q \longrightarrow \Sigma_j f_j \cdot S_j(q)$, has as image precisely $\mathcal{S}|U$. It is evident that, restricting U to be sufficiently small, there is no loss of generality in supposing that no germ $S_i(p) \in \mathcal{O}^m_p$ can be written as a linear combination of the remaining elements, with coefficients in the ring $\mathcal{O}_p$. The kernel $\mathcal{K}$ of the homomorphism φ then has as its stalk at p the zero stalk. (For if $(f_1,\ldots,f_r) \in \mathcal{K}_p$ and if $(f_1,\ldots,f_r) \neq (0,\ldots,0)$, let f_i be the element having a zero at p of least order; thus $f_j/f_i \in \mathcal{O}_p$ for all j. But then $0 = \varphi(f_1,\ldots,f_r) = \Sigma_j f_j \cdot S_j(p)$, and hence $S_i(p) = -\sum_{j\neq i} (f_j/f_i)S_j(p)$, a contradiction.) The kernel $\mathcal{K}$ is also coherent, so will be generated over U (restricting U further if necessary) by some sections $F_i \in \Gamma(U, \mathcal{O}^r)$;

but since $\mathcal{K}_p = 0$, necessarily all the germs $F_i(p) = (0,\ldots,0)$, and hence $\mathcal{K} = 0$ throughout U . Thus $\varphi\colon \mathcal{O}^r|U \longrightarrow \mathcal{S}|U$ is an isomorphism, and the proof is thereby concluded.

Corollary. If $\mathcal{S}$ is a coherent analytic sheaf over a Riemann surface M , then each point $p \in M$ has an open neighborhood U over which there is an exact sequence of analytic sheaves of the form

$$(1) \qquad 0 \longrightarrow \mathcal{O}^{m_1}|U \xrightarrow{\varphi_1} \mathcal{O}^{m}|U \xrightarrow{\varphi} \mathcal{S}|U \longrightarrow 0 .$$

Moreover, if U is a coordinate neighborhood, then whenever there is an exact sequence of the form (1),

$$H^q(U, \mathcal{S}) = 0 \quad \text{for all} \quad q > 0 ;$$

and corresponding to the sheaf sequence (1) is the exact sequence of sections

$$(2) \qquad 0 \longrightarrow \Gamma(U, \mathcal{O}^{m_1}) \xrightarrow{\varphi_1} \Gamma(U, \mathcal{O}^{m}) \xrightarrow{\varphi} \Gamma(U, \mathcal{S}) \longrightarrow 0 .$$

Proof. Since $\mathcal{S}$ is coherent, each point $p \in M$ has an open neighborhood U for which there is a sheaf homomorphism $\varphi\colon \mathcal{O}^m|U \longrightarrow \mathcal{S}|U$ which is onto; and the kernel of φ is coherent. The kernel is then locally free by Theorem 4, so that there is an exact sheaf sequence of the form (1) if U is sufficiently small. Recall from last year's lectures (Corollary to Theorem 4, page 44) that for a coordinate neighborhood U , $H^q(U, \mathcal{O}) = 0$ for all $q > 0$; but then the exact cohomology sequence corresponding to (1) begins with the sequence (2), and for the higher terms, all vanish except perhaps $H^q(U, \mathcal{S})$. This suffices to conclude the proof.

With a little more effort, and a few more preliminaries, it is possible to establish a semi-local form of the exact sequence (1).

(b) Let M be an arbitrary Riemann surface, and consider an open subset $U \subset M$. The bounded holomorphic functions form a linear subspace $\Gamma_0(U, \mathcal{O}) \subset \Gamma(U, \mathcal{O})$ of the space of all holomorphic functions on U. For any element $f \in \Gamma_0(U, \mathcal{O})$ put

$$\|f\| = \sup_{p \in U} |f(p)| ;$$

this is clearly a norm on the vector space $\Gamma_0(U, \mathcal{O})$, under which $\Gamma_0(U, \mathcal{O})$ becomes a Banach space. (To verify the latter assertion it is only necessary to show that $\Gamma_0(U, \mathcal{O})$ is complete in norm; but this is obvious, since convergence in norm is equivalent to uniform convergence on U.) The spaces $\Gamma_0(U, \mathcal{O}^n) = \Gamma_0(U, \mathcal{O})^n$ can be given a corresponding Banach space structure, defining for any element $F = (f_1, \ldots, f_n) \in \Gamma_0(U, \mathcal{O}^n)$ the norm $\|F\| = \max_j \|f_j\|$. Of particular interest will be the space $\Gamma_0(U, \mathcal{O}^{m\times m})$ of bounded holomorphic $m \times m$-matrix valued functions, which can be identified with the space $\Gamma_0(U, \mathcal{O}^{m^2})$. (The vector space of all complex $m \times m$ matrices will be denoted by $\mathbb{C}^{m\times m}$, and can be identified with $\mathbb{C}^{m^2}$; and $\Gamma_0(U, \mathcal{O}^{m\times m})$ is the set of bounded, complex analytic mappings from U into $\mathbb{C}^{m\times m}$.) If U, U_1 are two open subsets of M and $U \subset U_1$, it is clear that the restriction to U of an element $f \in \Gamma_0(U_1, \mathcal{O})$ is an element $\rho_U(f) \in \Gamma_0(U, \mathcal{O})$; indeed, the restriction mapping $\rho_U: \Gamma_0(U_1, \mathcal{O}) \longrightarrow \Gamma_0(U, \mathcal{O})$ is a continuous linear mapping between these Banach spaces.

The notation ρ_U will also be used for the restriction mapping between other spaces of analytic functions. Thus, for instance, $\rho_U\colon GL(m, \mathcal{O}_{U_1}) \longrightarrow GL(m, \mathcal{O}_U)$ is a group homomorphism, where as before $GL(m, \mathcal{O}_U) = \Gamma(U, \mathcal{GL}(m, \mathcal{O}))$.

<u>Theorem 4.</u> Let U_1, U_2 be open subsets of a Riemann surface M, with intersection $U = U_1 \cap U_2$; and assume that the linear mapping

$$\Theta\colon \Gamma_0(U_1, \mathcal{O}) \oplus \Gamma_0(U_2, \mathcal{O}) \longrightarrow \Gamma_0(U, \mathcal{O}),$$

defined by $\Theta(f_1, f_2) = \rho_U(f_1) + \rho_U(f_2)$, is onto. Let

$$\Phi\colon GL(m, \mathcal{O}_{U_1}) \times GL(m, \mathcal{O}_{U_2}) \longrightarrow GL(m, \mathcal{O}_U)$$

be defined by $\Phi(F_1, F_2) = \rho_U(F_1)\cdot\rho_U(F_2)$. Then there is an open neighborhood Δ of the identity element $I \in GL(m, \mathbb{C})$ such that the image of Φ contains at least the set

$$\{F \in GL(m, \mathcal{O}_U) \mid F(p) \in \Delta \text{ for all } p \in U\}.$$

Proof. Recall that the matrix exponential mapping $\exp\colon \mathbb{C}^{m\times m} \longrightarrow GL(m, \mathbb{C})$ defines a complex analytic homeomorphism between an open neighborhood D_0 of the origin in $\mathbb{C}^{m\times m}$ and an open neighborhood Δ_0 of the identity $I \in GL(m, \mathbb{C})$; let $\exp^{-1}\colon \Delta_0 \longrightarrow D_0$ be its inverse. (For the definitions and elementary properties of the matrix exponential function, see for instance C. Chevalley, <u>Theory of Lie Groups</u>, I, (Princeton Univ. Press, 1946), especially Chapter I.) Let $D_1, D_2 \subset D_0$ be open neighborhoods of the origin in $\mathbb{C}^{m\times m}$ such that $\exp X_1 \cdot \exp X_2 \in \Delta_0$ whenever $X_j \in D_j$.

Let $\Omega_j \subset \Gamma_0(U_j, \mathcal{O}^{m\times m})$ be the open subset of that Banach space defined by

$$\Omega_j = \{G \in \Gamma_0(U_j, \mathcal{O}^{m\times m}) | G(p) \in D_j \quad \text{for all} \quad p \in U_j\} .$$

It is then possible to define a mapping

$$\Psi\colon \Omega_1 \oplus \Omega_2 \longrightarrow \Gamma_0(U, \mathcal{O}^{m\times m})$$

by putting

$$\Psi(G_1, G_2) = \exp^{-1}(\exp G_1 \cdot \exp G_2) \ ;$$

and it is evident that Ψ is a continuous mapping from the open subset $\Omega_1 \oplus \Omega_2 \subset \Gamma_0(U_1, \mathcal{O}) \oplus \Gamma_0(U_2, \mathcal{O})$ into the Banach space $\Gamma_0(U, \mathcal{O}^{m\times m})$. To prove the desired theorem, it is sufficient to show that the image of the mapping Ψ contains an open neighborhood of the origin in $\Gamma_0(U, \mathcal{O}^{m\times m})$. For the image of Ψ would then contain a basic open neighborhood $U \subset \Gamma_0(U, \mathcal{O}^{m\times m})$ of the form

$$\Omega = \{G \in \Gamma_0(U, \mathcal{O}^{m\times m}) | G(p) \in D \quad \text{for all} \quad p \in U\} ,$$

where $D \subset D_0$ is an open subneighborhood of the origin. Letting $\Delta = \exp(D) \subset GL(m, \mathbf{C})$, whenever $F \in GL(m, \mathcal{O}_U)$ is such that $F(p) \in \Delta$ for all $p \in U$, there is an element $G \in \Omega$ such that $F = \exp G$; but then $G = \Psi(G_1, G_2)$ for some elements $G_j \in \Omega_j$, and putting $F_j = \exp G_j \in \Gamma(U_j, \mathcal{GL}(m, \mathcal{O})) = GL(m, \mathcal{O}_{U_j})$, it follows that $F = \exp \Psi(G_1, G_2) = \Phi(F_1, F_2)$, as desired. The proof will be completed by establishing some properties of the mapping Ψ , and using some general results about mappings between Banach spaces.

Let $\psi: D_1 \times D_2 \longrightarrow D_0 \subset \mathbb{C}^{m\times m}$ be the complex analytic mapping defined by $\psi(Z_1,Z_2) = \exp^{-1}(\exp Z_1 \cdot \exp Z_2)$. In the Taylor series expansion of this function at the origin $(0,0) \in D_1 \times D_2$, the constant term is the matrix $\psi(0,0) = 0$. The first-order terms can be viewed as a linear mapping $\lambda: \mathbb{C}^{m\times m} \oplus \mathbb{C}^{m\times m} \longrightarrow \mathbb{C}^{m\times m}$; and recalling the series expansion for the matrix exponential function, note that

$$\exp \psi(Z_1,Z_2) = I + \lambda(Z_1,Z_2) + (\text{higher order terms})$$

and that

$$\begin{aligned}\exp \psi(Z_1,Z_2) &= \exp Z_1 \cdot \exp Z_2 \\ &= I + (Z_1 + Z_2) + (\text{higher order terms}),\end{aligned}$$

hence that $\lambda(Z_1,Z_2) = Z_1 + Z_2$. Writing $|Z| = \max_{i,j}|z_{ij}|$ for any matrix $Z = (z_{ij}) \in \mathbb{C}^{m\times m}$, and $|(Z_1,Z_2)| = \max(|Z_1|,|Z_2|)$ for any matrix pair $(Z_1,Z_2) \in \mathbb{C}^{m\times m} \oplus \mathbb{C}^{m\times m}$, note then that

$$(3) \qquad \lim_{(Z_1,Z_2)\to(0,0)} \frac{|\psi(Z_1,Z_2) - \psi(0,0) - \lambda(Z_1,Z_2)|}{|(Z_1,Z_2)|} = 0 .$$

Defining then the linear mapping

$$\Lambda: \Gamma_0(U_1, \mathcal{O}^{m\times m}) \times \Gamma_0(U_2, \mathcal{O}^{m\times m}) \longrightarrow \Gamma_0(U, \mathcal{O}^{m\times m})$$

by $\Lambda(G_1,G_2) = \lambda(\rho_U G_1, \rho_U G_2) = \rho_U G_1 + \rho_U G_2$, it follows directly from (3) that

$$\lim_{(G_1,G_2)\to(0,0)} \frac{\|\Psi(G_1,G_2) - \Psi(0,0) - \Lambda(G_1,G_2)\|}{\|(G_1,G_2)\|} = 0 .$$

The function Ψ is therefore differentiable at the origin, and its derivative is $d\Psi(0,0) = \Lambda$. The same argument, with its obvious modifications, shows that Ψ is indeed strictly differentiable. The derivative $d\Psi(0,0) = \Lambda$, on the other hand, can be identified with the direct sum of m^2 copies of the mapping Θ ; and by hypothesis it is necessarily onto. This suffices to show that the image $\Psi(\Omega_1 \oplus \Omega_2)$ contains an open neighborhood of the origin, and hence to conclude the proof, in view of the following general result.

Lemma 5. Let E and F be Banach spaces, $\Omega \subset E$ be an open neighborhood of the origin, and $\Psi: \Omega \longrightarrow F$ be a continuous mapping. If Ψ is strictly differentiable at the origin, and if its derivative $d\Psi(0): E \longrightarrow F$ is onto, then the image $\Psi(\Omega) \subset F$ contains an open neighborhood of the image $\Psi(0)$.

Remarks. Before turning to the proof of the lemma, it might be convenient to recall the relevant definitions. A continuous mapping $\Psi: \Omega \longrightarrow F$ is said to be differentiable at the origin if there is a continuous linear mapping $\Lambda: E \longrightarrow F$ such that

$$\lim_{x \to 0} \frac{\|\Psi(x) - \Psi(0) - \Lambda(x)\|}{\|x\|} = 0 ;$$

the mapping Λ , which is evidently unique, is called the derivative of Ψ at the origin, and is denoted by $\Lambda = d\Psi(0)$. The mapping Ψ is said to be strictly differentiable at the origin if it is differentiable with derivative Λ and if moreover

$$\lim_{x_1, x_2 \to 0} \frac{\|\Psi(x_1) - \Psi(x_2) - \Lambda(x_1 - x_2)\|}{\|x_1 - x_2\|} = 0 .$$

Proof. To simplify the notation, suppose that $\Psi(0) = 0$. Since the continuous linear mapping $\Lambda = d\Psi(0)$ is onto, it follows from the open mapping theorem (see Dunford-Schwartz, Linear Operators I, (Interscience, New York, 1964), pages 55 ff.), that the image under Λ of the open ball B_1 of radius 1 centered at the origin contains an open ball B_c of some radius $c > 0$ centered at the origin in F; hence for any element $y \in F$ there exists an element $x \in E$ such that $\Lambda(x) = y$ and $\|x\| < \frac{2}{c}\|y\|$. Letting $\Psi_0 = \Psi - \Lambda$, it follows immediately from the strict differentiability of Ψ that for a sufficiently small constant $r > 0$,

$$\|\Psi_0(x_1) - \Psi_0(x_2)\| = \|\Psi(x_1) - \Psi(x_2) - \Lambda(x_1 - x_2)\| < \frac{c}{4}\|x_1 - x_2\|$$

whenever $x_1, x_2 \in B_r \subset \Omega$.

The proof will be concluded by showing that $\Psi(B_r) \supset B_{cr/4}$. For any point $y \in B_{cr/4}$, there exists a sequence of points $x_n \in B_r$ such that the following conditions hold:

(i) $x_0 = 0$;

(ii) $\Lambda(x_{n+1}) = y - \Psi_0(x_n)$;

(iii) $\|x_{n+1} - x_n\| < \frac{1}{2^{n-1}c}\|y\|$.

(To see this, note that x_1 must be so chosen that $\Lambda(x_1) = y$; but as above, there exists such a point with $\|x_1\| < \frac{2}{c}\|y\|$. Having obtained the points $x_1,\ldots,x_{n+1}$, the next point x_{n+2} must be so selected that $\Lambda(x_{n+2}) = y - \Psi_0(x_n)$; the difference $x_{n+2} - x_{n+1} = y_{n+2}$ must then be such that $\Lambda(y_{n+2}) = \Lambda(x_{n+2}) - \Lambda(x_{n+1}) = - \Psi_0(x_{n+1}) + \Psi_0(x_n)$. Note that $\|\Lambda(y_{n+2})\| = \|\Psi_0(x_{n+1}) - \Psi_0(x_n)\| < \frac{c}{4}\|x_{n+1} - x_n\| < \frac{1}{2^{n+1}}\|y\|$; but there exists such a point y_{n+2} for which $\|y_{n+2}\| < \frac{2}{c} \cdot \frac{1}{2^{n+1}} \|y\| = \frac{1}{2^n c} \|y\|$, and $x_{n+2} = x_{n+1} + y_{n+2}$ is the desired next element.) It should be observed by the way that $\|x_{n+1}\| \leq \|x_{n+1} - x_n\| + \ldots + \|x_2 - x_1\| + \|x_1 - x_0\| < \frac{4}{c} \|y\|$, so that all these points do lie in $B_r \subset \Omega$. It follows from (iii) that the sequence of points x_n converge to a limit point $x \in E$; and note that $\|x\| \leqq \frac{4}{c} \|y\| < r$, so that $x \in B_r$. But it follows from (ii) that then $\Lambda(x) = y - \Psi_0(x)$, so that $\Psi(x) = \Lambda(x) + \Psi_0(x) = y$; thus $y \in \Psi(B_r)$, and the proof is thereby concluded.

Remarks. The demonstration of Theorem 4 given here is essentially due to Malgrange and Grauert; compare also the notes of the Séminaire d'Analyse (Séminaire Frenkel), Strasbourg, 1964-1965. The first statement of a theorem of this sort, with quite a different proof, was due to H. Cartan, (J. Math. pures et appl. 19 (1940), 1-26). The statement and proof given here carry over almost verbatim for several complex variables.

Lemma 6. Let U_1, U_2 be bounded open subsets of the complex line $\mathbb{C}$, with intersection $U = U_1 \cap U_2$; and suppose that $\overline{U_1 - U} \cap \overline{U_2 - U} = \emptyset$. Then the mapping

$$\Theta\colon \Gamma_0(U_1, \mathcal{O}) \oplus \Gamma_0(U_2, \mathcal{O}) \longrightarrow \Gamma_0(U, \mathcal{O}),$$

defined by $\Theta(f_1, f_2) = \rho_U(f_1) + \rho_U(f_2)$, is onto.

Proof. Since $\overline{U_1 - U}$ and $\overline{U_2 - U}$ are disjoint closed subsets of $\mathbb{C}$, there is a C^∞ real-valued function $r(z)$ in $\mathbb{C}$ such that $r(z) \equiv 0$ in an open neighborhood of $\overline{U_1 - U}$ and $r(z) \equiv 1$ in an open neighborhood of $\overline{U_2 - U}$. If $f \in \Gamma_0(U, \mathcal{O})$ is any bounded holomorphic function, set

$$g_1(z) = \begin{cases} r(z)f(z) & \text{if } z \in U \\ 0 & \text{if } z \in U_1 - U \end{cases} ; \quad g_2(z) = \begin{cases} (1-r(z))f(z) & \text{if } z \in U \\ 0 & \text{if } z \in U_2 - U. \end{cases}$$

It is then clear that $g_j(z)$ is a C^∞ bounded function in U_j, and that $\rho_U(g_1) + \rho_U(g_2) = f$. The differential form $\varphi(z) \in \Gamma(U_1 \cup U_2, \mathcal{E}^{0,1})$ defined by

$$\varphi(z) = \begin{cases} \bar\partial g_1(z) & \text{if } z \in U_1 \\ -\bar\partial g_2(z) & \text{if } z \in U_2 \end{cases}$$

is then a well-defined C^∞ differential form. It was proved in last year's lectures that there is a C^∞ function $h(z)$ in $U_1 \cup U_2$ such that $\bar\partial h = \varphi$, (Theorem 4, page 42); and by examining the proof of that theorem, it is evident that h can be chosen to be bounded if φ is bounded, (but see the remarks at the end of this proof). The functions $f_1(z) = g_1(z) - h(z)$ for $z \in U_1$, and $f_2(z) = g_2(z) + h(z)$ for $z \in U_2$, are clearly so defined that $f_j \in \Gamma_0(U_j, \mathcal{O})$ and $\Theta(f_1, f_2) = f$, thus concluding the proof.

Remarks. The proof of Theorem 4 of last year's lectures was complicated by the necessity of allowing for unbounded differential forms; for the case of a bounded differential form φ, the desired function h can be taken simply as

$$h(z) = \frac{-1}{2\pi i} \iint_{U_1 \cup U_2} \frac{1}{\zeta - z} \psi(\zeta) \wedge d\zeta .$$

Lemma 7. Let V be a simply-connected open subset of the complex line $\mathbb{C}$, and let $K \subset V$ be a compact subset of V. Then any non-singular holomorphic matrix-valued function $F \in GL(m, \mathcal{O}_V)$ can be uniformly approximated on K by an element $H \in GL(m, \mathcal{O}_{\mathbb{C}})$.

Proof. Again let $D \subset \mathbb{C}^{m \times m}$ be an open neighborhood of the origin such that the matrix exponential function establishes an analytic homeomorphism $\exp: D \longrightarrow \Delta$, where Δ is an open neighborhood of the identity $I \in GL(m, \mathbb{C})$. First, suppose that $F \in GL(m, \mathcal{O}_V)$ has the property that $F(p) \in \Delta$ for all points p in an open neighborhood of K; thus $F = \exp G$ for some function $G \in \mathcal{O}_W^{m \times m}$, for some open neighborhood W of K. By the ordinary Runge theorem, the function G can be uniformly approximated on K by a polynomial P; and then $H = \exp P$ is a non-singular holomorphic matrix-valued function in $\mathbb{C}$ which approximates F uniformly on K. Next, for an arbitrary function $F \in GL(m, \mathcal{O}_V)$, there will exist finitely many functions $F_i \in GL(m, \mathcal{O}_V)$ such that $F_i(p) \in \Delta$ for all points p in an open neighborhood of K, and such that $F = F_1 \cdot \ldots \cdot F_n$. (For the set $GL(m, \mathcal{O}_V)$ is a topological group with the compact-open topology; it is readily noted that the group is connected, hence it is generated by an open neighborhood of the identity. One such open neighborhood consists of all elements $F_0 \in GL(m, \mathcal{O}_V)$ such that $F_0(p) \in \Delta$ for all points p in a compact set containing K in its interior; and thus any element $F \in GL(m, \mathcal{O}_V)$ is a finite product of elements from this open subset.) Applying the preceding argument to each function F_i leads to the desired result, thus concluding the proof.

<u>Theorem 5.</u> Let $D \subset \mathbf{C}$ be an open subset of the complex plane such that its closure $\bar{D}$ is compact and simply connected; and let $\mathcal{S}$ be a coherent analytic sheaf over an open neighborhood of $\bar{D}$. Then there is an exact sequence of analytic sheaves of the form

$$(4) \qquad 0 \longrightarrow \mathcal{O}^{m_1}|D \xrightarrow{\varphi_1} \mathcal{O}^{m}|D \xrightarrow{\varphi} \mathcal{S}\,|D \longrightarrow 0 .$$

Proof. By the Riemann mapping theorem, there is no loss of generality in supposing that D is a bounded rectangular subdomain of $\mathbf{C}$. The corollary to Theorem 3 shows that there is an exact sequence of the form (4) over an open neighborhood of each point $p \in \bar{D}$. Thus D can be decomposed by a sufficiently fine rectangular grating (that is, by a finite number of lines parallel to the real or imaginary axis) into smaller rectangular segments, on an open neighborhood of each of which there is an exact sequence of the desired form. To complete the proof, it is merely necessary to describe how to patch together exact sequences over two neighboring rectangles into an exact sequence over their union; for this process can be used to patch together the sequences in each horizontal row first, and then to patch together the various rows into an exact sequence over D.

Thus suppose that U_1, U_2 are open rectangular neighborhoods of two adjacent rectangles, as in the following diagram, and let $U = U_1 \cap U_2$.

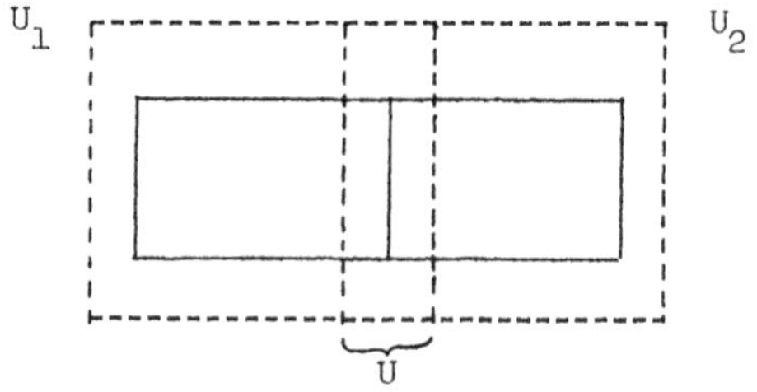

Suppose further that over an open neighborhood V_j of $\overline{U}_j$ there is an exact sequence of analytic sheaves of the form

$$0 \longrightarrow \mathcal{O}^{r_j}|V_j \xrightarrow{\rho_j} \mathcal{O}^{s_j}|V_j \xrightarrow{\sigma_j} \mathcal{S}|V_j \longrightarrow 0 .$$

There are thus two such sequences over the intersection $V = V_1 \cap V_2$; and it is clear that there are sheaf homomorphisms α, β , such that the following diagram is commutative, where <u>id</u> denotes the identity homomorphism.

$$\begin{array}{ccccccccc}
0 & \longrightarrow & \mathcal{O}^{r_1}|V & \xrightarrow{\rho_1} & \mathcal{O}^{s_1}|V & \xrightarrow{\sigma_1} & \mathcal{S}|V & \longrightarrow & 0 \\
 & & & & \alpha \downarrow \; \uparrow \beta & & \downarrow \uparrow \, id & & \\
0 & \longrightarrow & \mathcal{O}^{r_2}|V & \xrightarrow{\rho_2} & \mathcal{O}^{s_2}|V & \xrightarrow{\sigma_2} & \mathcal{S}|V & \longrightarrow & 0
\end{array}$$

(To see this, let $E_i \in \Gamma(V, \mathcal{O}^{s_1})$ be the canonical generating sections, and put $S_i = \sigma_1(E_i) \in \Gamma(V, \mathcal{S})$. As in the corollary to Theorem 4, there are sections $F_i \in \Gamma(V, \mathcal{O}^{s_2})$ such that $\sigma_2(F_i) = S_i$. Letting α be the homomorphism defined by $\alpha(E_i) = F_i$, it is evident that $\sigma_2\alpha = \sigma_1$, as desired. The mapping β is constructed similarly.) Now define sheaf homomorphisms

$$\theta\colon\ \mathcal{O}^{s_1} \oplus \mathcal{O}^{s_2} \longrightarrow \mathcal{O}^{s_2} \oplus \mathcal{O}^{s_1} , \qquad \theta'\colon\ \mathcal{O}^{s_2} \oplus \mathcal{O}^{s_1} \longrightarrow \mathcal{O}^{s_1} \oplus \mathcal{O}^{s_2}$$

by

$$\theta(F,G) = (G + \alpha F - \alpha\beta G,\ F - \beta G)$$
$$\theta'(K,L) = (L + \beta K - \beta\alpha L,\ K - \alpha L) .$$

It is readily verified that $\theta'\theta = \theta\theta' =$ identity, and hence that θ is an isomorphism with θ' as its inverse; and the following diagram of exact sequences is also commutative.

$$\begin{array}{ccccccccc} 0 \longrightarrow & \mathcal{O}^{r_1} \oplus \mathcal{O}^{s_2}|V & \xrightarrow{(\rho_1, id)} & \mathcal{O}^{s_1} \oplus \mathcal{O}^{s_2}|V & \xrightarrow{\sigma_1} & \mathcal{S}|V & \longrightarrow 0 \\ & & & \downarrow \theta & & \downarrow id & \\ 0 \longrightarrow & \mathcal{O}^{r_2} \oplus \mathcal{O}^{s_1}|V & \xrightarrow{(\rho_2, id)} & \mathcal{O}^{s_2} \oplus \mathcal{O}^{s_1}|V & \xrightarrow{\sigma_2} & \mathcal{S}|V & \longrightarrow 0 \end{array}$$

The mapping θ is defined by a non-singular holomorphic matrix-valued function over V, that is, by an element $\Theta \in GL(s_1 + s_2, \mathcal{O}_V)$. By Lemma 7, the matrix Θ can be approximated uniformly over the compact subset $\overline{U} \subset V$ by a non-singular holomorphic matrix-valued function over $V_1 \cup V_2$; thus there is an element $\Theta_0 \in GL(s_1 + s_2, \mathcal{O}_{V_1 \cup V_2})$ such that for all points $p \in \overline{U}$ the matrix $\Theta(p)\Theta_0(p)^{-1} \in \Delta$, where Δ is an open neighborhood of the identity $I \in GL(s_1 + s_2, \mathbb{C})$. In particular, select Δ sufficiently small that Theorem 4 applies. Then, recalling by Lemma 6 that all the hypotheses of Theorem 4 are fulfilled, it follows that there are elements $\Theta_1' \in GL(s_1 + s_2, \mathcal{O}_{U_1})$ and $\Theta_2 \in GL(s_1 + s_2, \mathcal{O}_{U_2})$ such that $\Theta(p)\cdot\Theta_0(p)^{-1} = \Theta_2(p)\cdot\Theta_1'(p)$ for all $p \in U$. Writing $\Theta_1 = \Theta_1'\Theta_0$, this condition can be written $\Theta(p) = \Theta_2(p)\cdot\Theta_1(p)$ for all $p \in U$. Letting θ_j be the sheaf isomorphisms defined by the matrices Θ_j, consider then the exact sequences

$$0 \longrightarrow \mathcal{O}^{r_1+s_2}|U_1 \longrightarrow \mathcal{O}^{s_1+s_2}|U_1 \xrightarrow{\sigma_1\theta_1^{-1}} \mathcal{S}|U_1 \longrightarrow 0$$

$$0 \longrightarrow \mathcal{O}^{r_2+s_1}|U_2 \longrightarrow \mathcal{O}^{s_2+s_1}|U_2 \xrightarrow{\sigma_2\theta_2} \mathcal{S}|U_2 \longrightarrow 0 .$$

Since on the subset U the construction yields the fact that $\sigma_1 = \sigma_2\theta = \sigma_2\theta_2\theta_1$, hence that $\sigma_1\theta_1^{-1} = \sigma_2\theta_2$, these two sequences coincide over U, and therefore determine an exact sequence of the desired form over $U_1 \cup U_2$. This then concludes the proof.

(c) Considering now the special case of the complex projective line $\mathbb{P}$, the structure of general coherent analytic sheaves can be described quite easily in terms of locally free analytic sheaves. To see this, recall that $\mathbb{P}$ has a standard complex analytic coordinate covering $\mathfrak{U} = \{(U_1,z_1),(U_2,z_2)\}$; here $U_j \subset \mathbb{P}$ are open subsets, $z_j\colon U_j \longrightarrow \mathbb{C}$ are homeomorphisms from U_j onto the entire complex line, the points $p \in U_1 \cap U_2$ are precisely those for which $z_1(p) \neq 0$ and $z_2(p) \neq 0$, and the coordinate transition functions are defined by $z_1(p) = 1/z_2(p)$ when $p \in U_1 \cap U_2$.

Theorem 6. If $\mathcal{S}$ is a coherent analytic sheaf over the projective line $\mathbb{P}$, then there are locally free analytic sheaves $\mathcal{F}$, $\mathcal{F}_1$ over $\mathbb{P}$ for which the following is an exact sequence of analytic sheaves.

$$0 \longrightarrow \mathcal{F}_1 \xrightarrow{\varphi_1} \mathcal{F} \xrightarrow{\varphi} \mathcal{S} \longrightarrow 0 .$$

Proof. In terms of the standard coordinate covering $\mathfrak{U}$ of the projective line, let $D_j \subset U_j$ be open subsets homeomorphic to the disc $\{z_j \in \mathbb{C} \mid |z_j| < 2\}$ under the coordinate mappings z_j ; the intersection $D = D_1 \cap D_2$ is thus a finite annulus in either coordinate system. By Theorem 5 there are exact sequences of analytic sheaves of the form

$$(5) \qquad 0 \longrightarrow \mathcal{O}^{r_j}|D_j \xrightarrow{\rho_j} \mathcal{O}^{s_j}|D_j \xrightarrow{\sigma_j} \mathcal{S}\,|D_j \longrightarrow 0 .$$

This then provides two such exact sequences over the intersection $D = D_1 \cap D_2$. Just as in the proof of Theorem 5, the sequences (5) can be so modified that over D there is an isomorphism $\theta\colon \mathcal{O}^{s_1}|D \to \mathcal{O}^{s_2}|D$ for which the following is a commutative diagram.

$$\begin{array}{ccccccccc}
0 & \longrightarrow & \mathcal{O}^{r_1}|D & \xrightarrow{\rho_1} & \mathcal{O}^{s_1}|D & \xrightarrow{\sigma_1} & \mathcal{S}\,|D & \longrightarrow & 0 \\
 & & \downarrow{\scriptstyle \theta_1} & & \downarrow{\scriptstyle \theta} & & \downarrow{\scriptstyle \mathrm{id}} & & \\
0 & \longrightarrow & \mathcal{O}^{r_2}|D & \xrightarrow{\rho_2} & \mathcal{O}^{s_2}|D & \xrightarrow{\sigma_2} & \mathcal{S}\,|D & \longrightarrow & 0
\end{array}$$

Note that necessarily $s_1 = s_2$ and $r_1 = r_2$; and that the restriction of θ to the subsheaf $\rho_1(\mathcal{O}^{r_1}|D) \subset \mathcal{O}^{s_1}|D$ defines an isomorphism $\theta_1\colon \mathcal{O}^{r_1}|D \to \mathcal{O}^{r_2}|D$, for which the diagram is still commutative. Now the elements θ and θ_1 can be represented by matrices $\Theta \in GL(s, \mathcal{O}_D)$ and $\Theta_1 \in GL(r, \mathcal{O}_D)$; and these matrices define complex fibre bundles for the covering $\{U_1, U_2\}$ or equivalently, locally free analytic sheaves $\mathcal{F}$, $\mathcal{F}_1$ respectively, over $\mathbb{P}$. By construction, these are just the sheaves desired to complete the proof of the theorem, however.

Having demonstrated this result, the basic existence theorem follows readily for the special case of the projective line, by a similar argument.

Theorem 7. Any vector bundle over the projective line $\mathbb{P}$ admits non-trivial meromorphic sections.

Proof. Let $\mathcal{U} = \{(U_1,z_1),(U_2,z_2)\}$ be the standard complex analytic coordinate covering of $\mathbb{P}$, and let $D_j \subset U_j$ be open subsets homeomorphic to the disc $\{z_j \in \mathbb{C} \mid |z_j| < 2\}$ under the coordinate mappings z_j. As in the proof of Theorem 6, a complex vector bundle $\Phi \in H^1(\mathbb{P}, \mathcal{GL}(m, \mathcal{O}))$ can be defined by a coordinate bundle for the covering $\{D_1,D_2\}$, hence by an element $\Phi_{12} \in GL(m, \mathcal{O}_D)$ where $D = D_1 \cap D_2$. In terms of the z_1-coordinate, for example, $\Phi_{12}(z_1)$ is a holomorphic, non-singular matrix-valued function in the annulus $D\colon \frac{1}{2} < |z_1| < 2$; and the Laurent expansions of the various entries yield a matrix Θ of rational functions on $\mathbb{P}$ which approximate Φ_{12} uniformly on any compact subset of the annulus D. (Recall that a rational function is a quotient of polynomial functions; of course, the only singularities of Θ are at the two points $z_1 = 0$ and $z_2 = 0$.) Thus there is a rational matrix Θ in $\mathbb{P}$, with singularities at most at the points $z_1 = 0$ and $z_2 = 0$, such that Θ is non-singular in the annulus $\tilde{D}\colon \frac{2}{3} < |z_j| < \frac{3}{2}$ and $\Phi_{12}(p)\Theta(p)^{-1} \in \Delta$ for all points $p \in \tilde{D}$, where Δ is an open neighborhood of the identity $I \in GL(m,\mathbb{C})$.

Let $\tilde{D}_j \subset \tilde{D}$ be the open subset homeomorphic to the disc $\{z_j \in \mathbb{C} \mid |z_j| < \frac{3}{2}\}$, so that $\tilde{D} = \tilde{D}_1 \cap \tilde{D}_2$. Note that any function $f \in \Gamma_0(\tilde{D}, \mathcal{O})$ can be written $f = f_1 + f_2$ for some functions $f_j \in \Gamma_0(\tilde{D}_j, \mathcal{O})$; this is an immediate consequence of the Laurent expansion of the function $f(z_j)$ in terms of either coordinate z_j.

Then, applying Theorem 4, there are holomorphic, non-singular matrix-valued functions $\Theta_j \in GL(m, \mathcal{O}_{\tilde{D}_j})$ such that $\Phi_{12}(p)\Theta(p)^{-1} = \Theta_1(p)\Theta_2(p)$ in $\tilde{D}$, hence such that

$$\Phi_{12}(p) = \Theta_1(p)\Theta_2(p)\Theta(p) \quad \text{for } p \in \tilde{D} .$$

Now, to construct a meromorphic section of the complex analytic vector bundle Φ, begin with any non-trivial (that is, not identically vanishing) meromorphic vector-valued function $F \in \Gamma(\mathbb{P}, \mathcal{M}^m)$. The functions $F_1 = \Theta_1 \cdot F \in \Gamma(\tilde{D}_1, \mathcal{M}^m)$ and $F_2 = \Theta^{-1}\Theta_2^{-1}F \in \Gamma(\tilde{D}_2, \mathcal{M}^m)$ are meromorphic in their respective neighborhoods; for Θ is not identically singular, and is rational on $\mathbb{P}$, so Θ^{-1} is meromorphic on $\mathbb{P}$. By construction then, $F_1(p) = \Theta_1(p)F(p) = \Phi_{12}(p)F_2(p)$ whenever $p \in \tilde{D}$, so that this is the desired meromorphic section, concluding the proof.

Remarks. It is apparent that the proof of Theorem 7 actually yields somewhat more, namely, that $H^1(\mathbb{P}, \mathcal{GL}(m, \mathcal{M})) = 0$, where $\mathcal{M}$ is again the sheaf of germs of meromorphic functions. (In other words, all complex meromorphic vector bundles over $\mathbb{P}$ are trivial, or all locally free sheaves of $\mathcal{M}$-modules over $\mathbb{P}$ are globally free.) Thus for any complex analytic vector bundle $\Phi \in H^1(\mathbb{P}, \mathcal{GL}(m, \mathcal{O}))$ it follows that $\mathcal{M}(\Phi) \cong \mathcal{M}^m$, and consequently, that $\Gamma(\mathbb{P}, \mathcal{M}(\Phi)) \cong \Gamma(\mathbb{P}, \mathcal{M}^m)$, the latter clearly being nonzero.

With these observations in mind, Theorem 6 can be extended as follows. Let $\mathcal{S}$ be any coherent analytic sheaf over the projective line $\mathbb{P}$, represented by an exact sequence of analytic sheaves of the form

$$0 \longrightarrow \mathcal{F}_1 \longrightarrow \mathcal{F} \longrightarrow \mathcal{S} \longrightarrow 0 ,$$

where $\mathcal{F}$ and $\mathcal{F}_1$ are locally free. Tensoring this exact sequence with the sheaf $\mathcal{M}$, considered as a sheaf of $\mathcal{O}$-modules, it follows readily from the condition that $\mathcal{F}$ and $\mathcal{F}_1$ are locally free that the following is also an exact sequence of analytic sheaves.

$$(6) \qquad 0 \longrightarrow \mathcal{M} \otimes_{\mathcal{O}} \mathcal{F}_1 \longrightarrow \mathcal{M} \otimes_{\mathcal{O}} \mathcal{F} \longrightarrow \mathcal{M} \otimes_{\mathcal{O}} \mathcal{S} \longrightarrow 0 .$$

If $\mathcal{F}$ is represented by the complex analytic vector bundle Φ, it is evident that $\mathcal{M} \otimes_{\mathcal{O}} \mathcal{F} = \mathcal{M} \otimes_{\mathcal{O}} \mathcal{O}(\Phi) = \mathcal{M}(\Phi) \cong \mathcal{M}^m$; so that (6) reduces to the exact sequence of sheaves of $\mathcal{M}$-modules

$$(7) \qquad 0 \longrightarrow \mathcal{M}^{m_1} \xrightarrow{\theta} \mathcal{M}^m \longrightarrow \mathcal{M} \otimes_{\mathcal{O}} \mathcal{S} \longrightarrow 0 .$$

The homomorphism θ in (7) is represented by an $m_1 \times m$ non-singular matrix Θ of rank m_1 over the field $\Gamma(\mathbb{P}, \mathcal{M})$; and after choosing coordinates such that this matrix has the form $\Theta = (\Theta_1, 0)$, where $\Theta_1 \in GL(m_1, \mathcal{M}_{\mathbb{P}})$, it follows from (7) that

$$\mathcal{M} \otimes_{\mathcal{O}} \mathcal{S} \cong \mathcal{M}^{m-m_1} .$$

Thus, <u>upon tensoring with</u> $\mathcal{M}$, <u>every coherent analytic sheaf over</u> $\mathbb{P}$ <u>reduces to a free sheaf of</u> $\mathcal{M}$<u>-modules.</u>

§3. Induced mappings of analytic sheaves

(a) Suppose that M and N are two arbitrary Riemann surfaces, with ${}_M\mathcal{O}$ and ${}_N\mathcal{O}$ as their respective sheaves of germs of holomorphic functions, and that $f: M \longrightarrow N$ is a non-trivial complex analytic mapping. (Here non-trivial means that M is not mapped to a single point. Recalling the discussion in §10(a) of last year's lectures, such a mapping is continuous and open, and exhibits M as a local branched covering of N .) Given a sheaf over one of the surfaces, the mapping f induces a sheaf over the other surface in a natural manner; these induced sheaves are the subject matter of the present section.

Consider firstly a given sheaf $\mathcal{S}$ of abelian groups over the Riemann surface N ; there is then an induced sheaf $\mathcal{R} = f^{-1}(\mathcal{S})$ of abelian groups over the Riemann surface M , defined as follows. Let $\mathcal{U} = \{U_\alpha\}$ be a basis for the topology of M , and to each set $U_\alpha \in \mathcal{U}$ associate the abelian group $\mathcal{R}_\alpha = \Gamma(f(U_\alpha), \mathcal{S})$; whenever $U_\alpha \subset U_\beta$, let $\rho_{\alpha\beta}: \mathcal{R}_\beta \longrightarrow \mathcal{R}_\alpha$ be the homomorphism defined by restricting an element of $\mathcal{R}_\beta = \Gamma(f(U_\beta), \mathcal{S})$ to the subset $f(U_\alpha) \subset f(U_\beta)$. It is evident that $\{U_\alpha, \mathcal{R}_\alpha, \rho_{\alpha\beta}\}$ is a presheaf of abelian groups over M ; and $\mathcal{R} = f^{-1}(\mathcal{S})$ is defined to be the associated sheaf. The same construction also applies to sheaves of rings, of course; so in particular, $f^{-1}({}_N\mathcal{O})$ is a well-defined sheaf of rings over the Riemann surface M . It should be noted that $f^{-1}({}_N\mathcal{O})$ can be identified in a natural manner as a subsheaf of ${}_M\mathcal{O}$. (For letting $\mathcal{R} = f^{-1}({}_N\mathcal{O})$ be defined by the presheaf $\{U_\alpha, \mathcal{R}_\alpha, \rho_{\alpha\beta}\}$ as above, to any element

$h_\alpha \in \mathcal{R}_\alpha = \Gamma(f(U_\alpha), {}_N\mathcal{O})$ associate the element $h_\alpha \circ f \in \Gamma(U_\alpha, {}_M\mathcal{O})$; this mapping clearly yields an isomorphism $f^{-1}({}_N\mathcal{O}) \longrightarrow {}_M\mathcal{O}$, as desired.) This identification will always be made, without further comment. Since $f^{-1}({}_N\mathcal{O}) \subset {}_M\mathcal{O}$ is a subsheaf of rings, ${}_M\mathcal{O}$ can be viewed as a sheaf of modules over the sheaf of rings $f^{-1}({}_N\mathcal{O})$.

This construction is not quite satisfactory for sheaves of modules over the sheaf of rings ${}_N\mathcal{O}$, that is, for analytic sheaves over the Riemann surface N. For letting $\mathcal{R} = f^{-1}(\mathcal{S})$ be defined by the presheaf $\{U_\alpha, \mathcal{R}_\alpha, \rho_{\alpha\beta}\}$ as above, each $\mathcal{R}_\alpha = \Gamma(f(U_\alpha), \mathcal{S})$ is a module over the ring $\Gamma(f(U_\alpha), {}_N\mathcal{O})$, where the latter rings form the presheaf leading to the sheaf $f^{-1}({}_N\mathcal{O})$; thus $\mathcal{R}$ has only the natural structure of a sheaf of modules over the sheaf of rings $f^{-1}({}_N\mathcal{O})$. Thus it is more natural in this context to consider the <u>inverse image sheaf</u> $f^{o}(\mathcal{S})$ over the Riemann surface M, defined by

(1) $$f^{o}(\mathcal{S}) = {}_M\mathcal{O} \otimes_{f^{-1}({}_N\mathcal{O})} f^{-1}(\mathcal{S}) .$$

This is then a sheaf of modules over the sheaf of rings ${}_M\mathcal{O}$, that is, an analytic sheaf over the Riemann surface M. Note in particular that $f^{o}({}_N\mathcal{O}) = {}_M\mathcal{O} \otimes_{f^{-1}({}_N\mathcal{O})} f^{-1}({}_N\mathcal{O}) = {}_M\mathcal{O}$.

The properties of this operation on sheaves are summarized in the following assertions.

<u>Lemma 8.</u> Let $f\colon M \longrightarrow N$ be a non-trivial complex analytic mapping between two Riemann surfaces, and let $\mathcal{S}, \mathcal{S}_1, \mathcal{S}_2, \ldots$ be analytic sheaves over the Riemann surface N.

(i) Then $f^{o}(\mathcal{S}_1 \oplus \mathcal{S}_2) = f^{o}(\mathcal{S}_1) \oplus f^{o}(\mathcal{S}_2)$, and

$$f^{o}(\mathcal{S}_1 \otimes_{{}_N\mathcal{O}} \mathcal{S}_2) = f^{o}(\mathcal{S}_1) \otimes_{{}_M\mathcal{O}} f^{o}(\mathcal{S}_2) .$$

(ii) If $\mathcal{S}_1 \longrightarrow \mathcal{S}_2 \longrightarrow \mathcal{S}_3 \longrightarrow 0$ is an exact sequence of ytic sheaves over N , then $\mathcal{S}_1) \longrightarrow f^{o}(\mathcal{S}_2) \longrightarrow f^{o}(\mathcal{S}_3) \longrightarrow 0$ is an exact sequence of ytic sheaves over M .

iii) If $\mathcal{S}$ is a locally free analytic sheaf of rank m over then $f^{o}(\mathcal{S})$ is a locally free analytic sheaf of rank m M .

iv) If $\mathcal{S}$ is a coherent analytic sheaf over N , then $f^{o}(\mathcal{S})$ coherent analytic sheaf over M .

Proof. (i) It is evident that $f^{-1}(\mathcal{S}_1 \oplus \mathcal{S}_2) =$ $^{1}(\mathcal{S}_1) \oplus f^{-1}(\mathcal{S}_2)$ and $f^{-1}(\mathcal{S}_1 \otimes_{{}_N\mathcal{O}} \mathcal{S}_2) =$ $^{1}(\mathcal{S}_1) \otimes_{f^{-1}({}_N\mathcal{O})} f^{-1}(\mathcal{S}_2)$; this can be seen by considering stalks at any point of the surface, since $(f^{-1}(\mathcal{S}))_p = \mathcal{S}_{f(p)}$.

$$f^{o}(\mathcal{S}_1 \oplus \mathcal{S}_2) = {}_M\mathcal{O} \otimes f^{-1}(\mathcal{S}_1 \oplus \mathcal{S}_2) = f^{o}(\mathcal{S}_1) \oplus f^{o}(\mathcal{S}_2) ;$$

recalling the standard properties of the tensor products of les,

$$f^{o}(\mathcal{S}_1 \otimes_{{}_M\mathcal{O}} \mathcal{S}_2) = {}_M\mathcal{O} \otimes_{f^{-1}({}_N\mathcal{O})} f^{-1}(\mathcal{S}_1) \otimes_{f^{-1}({}_N\mathcal{O})} f^{-1}(\mathcal{S}_2)$$

$$\cong f^{-1}(\mathcal{S}_1) \otimes_{f^{-1}({}_N\mathcal{O})} {}_M\mathcal{O} \otimes_{f^{-1}({}_N\mathcal{O})} f^{-1}(\mathcal{S}_2)$$

$$\cong f^{-1}(\mathcal{S}_1) \otimes_{f^{-1}({}_N\mathcal{O})} ({}_M\mathcal{O} \otimes_{{}_M\mathcal{O}} {}_M\mathcal{O}) \otimes_{f^{-1}({}_N\mathcal{O})} f^{-1}(\mathcal{S}_2)$$

$$\cong (f^{-1}(\mathcal{S}_1) \otimes_{f^{-1}({}_N\mathcal{O})} {}_M\mathcal{O}) \otimes_{{}_M\mathcal{O}} ({}_M\mathcal{O} \otimes_{f^{-1}({}_N\mathcal{O})} f^{-1}(\mathcal{S}_2))$$

$$\cong f^{o}(\mathcal{S}_1) \otimes_{{}_M\mathcal{O}} f^{o}(\mathcal{S}_2) .$$

(ii) It is an immediate consequence of the hypotheses that $f^{-1}(\mathcal{S}_1) \longrightarrow f^{-1}(\mathcal{S}_2) \longrightarrow f^{-1}(\mathcal{S}_3) \longrightarrow 0$ is an exact sequence of sheaves over M ; tensoring with ${}_M\mathcal{O}$, and recalling the properties of the tensor product (especially page 4), it follows that ${}_M\mathcal{O} \otimes f^{-1}(\mathcal{S}_1) \longrightarrow {}_M\mathcal{O} \otimes f^{-1}(\mathcal{S}_2) \longrightarrow {}_M\mathcal{O} \otimes f^{-1}(\mathcal{S}_3) \to 0$ is exact as desired.

(iii) As noted earlier, $f^{o}({}_N\mathcal{O}) = {}_M\mathcal{O}$; and it follows from this and part (a) above that $f^{o}({}_N\mathcal{O}^{m}) = {}_M\mathcal{O}^{m}$. Now suppose that $\mathcal{S}$ is locally free of rank m over N , and consider a point $p \in M$. Since f is continuous, there is an open neighborhood U of p sufficiently small that $\mathcal{S}|f(U) \cong {}_N\mathcal{O}^{m}|f(U)$; but then $f^{o}(\mathcal{S})|U = f^{o}(\mathcal{S}|f(U)) \cong f^{o}({}_N\mathcal{O}^{m}|f(U)) = {}_M\mathcal{O}^{m}$, so that $f^{o}(\mathcal{S})$ is locally free of rank m over M .

(iv) Finally, suppose that $\mathcal{S}$ is a coherent analytic sheaf over N , and consider a point $p \in M$. Again, by the continuity of f_1 there is an open neighborhood U of p with image as small as desired, in particular, so small that there is an exact sequence of analytic sheaves ${}_N\mathcal{O}^{m_1}|f(U) \longrightarrow {}_N\mathcal{O}^{m}|f(U) \longrightarrow \mathcal{S}|f(U)$. Then, using parts (b) and (c), there is an exact sequence of analytic sheaves over U of the form

${}_M\mathcal{O}^{m_1}|U \longrightarrow {}_M\mathcal{O}^{m}|U \longrightarrow f^o(\mathcal{S})|U \longrightarrow 0$, so that $f^o(\mathcal{S})$ is a coherent analytic sheaf over M , completing the proof of the lemma.

(b) Consider next a given sheaf $\mathcal{S}$ of abelian groups over the Riemann surface M , and assume that the mapping $f: M \longrightarrow N$ is onto all of N ; there is then an induced sheaf $\mathcal{J} = f_o(\mathcal{S})$ of abelian groups over the Riemann surface N , called the <u>direct image</u> of the sheaf $\mathcal{S}$, defined as follows. Let $\mathcal{U} = \{U_\alpha\}$ be a basis for the topology of N , and to each set $U_\alpha \in \mathcal{U}$ associate the abelian group $\mathcal{J}_\alpha = \Gamma(f^{-1}(U_\alpha), \mathcal{S})$; whenever $U_\alpha \subset U_\beta$, let $\rho_{\alpha\beta}: \mathcal{J}_\beta \longrightarrow \mathcal{J}_\alpha$ be the homomorphism defined by restricting an element of $\Gamma(f^{-1}(U_\beta), \mathcal{S})$ to the subset $f^{-1}(U_\alpha) \subset f^{-1}(U_\beta)$. It is evident that $\{U_\alpha, \mathcal{J}_\alpha, \rho_{\alpha\beta}\}$ is a presheaf of abelian groups over N ; and $\mathcal{J} = f_o(\mathcal{S})$ is defined to be the associated sheaf. Note that this presheaf is a complete presheaf. (For suppose that U_o and $\{U_\beta\}$ are sets of the covering $\mathcal{U}$ such that $U_o = \cup_\beta U_\beta$. Firstly, if $h_o \in \mathcal{J}_o = \Gamma(f^{-1}(U_o), \mathcal{S})$ is an element such that $\rho_{\beta o}(h_o) = 0$ for all β , then h_o is the zero section over $f^{-1}(U_\beta)$, and hence necessarily $h_o = 0$. Secondly, if $h_\beta \in \mathcal{J}_\beta = \Gamma(f^{-1}(U_\beta), \mathcal{S})$ are elements such that $\rho_{\gamma\beta_1}(h_{\beta_1}) = \rho_{\gamma\beta_2}(h_{\beta_2})$ whenever $U_\gamma \subset U_{\beta_1} \cap U_{\beta_2}$, then the sections h_{β_1}, h_{β_2} agree over $f^{-1}(U_{\beta_1} \cap U_{\beta_2}) = f^{-1}(U_{\beta_1}) \cap f^{-1}(U_{\beta_2})$; there is therefore a section

$h_o \in \mathcal{J}_o = \Gamma(f^{-1}(U_o), \mathcal{S})$ such that $\rho_{\beta o}(h_o) = h_\beta$ for all β .) As a consequence, there are natural identifications $\Gamma(U_\alpha, f_o(\mathcal{S})) = \mathcal{J}_\alpha = \Gamma(f^{-1}(U_\alpha), \mathcal{S})$. In particular, there is a natural isomorphism

$$(2) \qquad \Gamma(N, f_o(\mathcal{S})) \cong \Gamma(M, \mathcal{S}) .$$

The same construction also applies to sheaves of rings, of course; so in particular, $f_o({}_M\mathcal{O})$ is a well-defined sheaf of rings over the Riemann surface N . It should be noted that ${}_N\mathcal{O}$ can be identified in a natural manner as a subsheaf of $f_o({}_M\mathcal{O})$. (For letting $\mathcal{J} = f_o({}_M\mathcal{O})$ be defined by the presheaf $\{U_\alpha, \mathcal{J}_\alpha, \rho_{\alpha\beta}\}$ as above, to any element $h_\alpha \in \Gamma(U_\alpha, {}_N\mathcal{O})$ associate the element $h_\alpha \circ f \in \Gamma(f^{-1}(U_\alpha), {}_M\mathcal{O}) = \mathcal{J}_\alpha$; this mapping clearly extends to an isomorphism ${}_N\mathcal{O} \longrightarrow f_o({}_M\mathcal{O})$, as desired.) This identification will always be made, without further comment. Since ${}_N\mathcal{O} \subset f_o({}_M\mathcal{O})$ is a subsheaf of rings, $f_o({}_M\mathcal{O})$ can be viewed as a sheaf of modules over the sheaf of rings ${}_N\mathcal{O}$.

If $\mathcal{S}$ is an analytic sheaf over M (a sheaf of ${}_M\mathcal{O}$ -modules), then the direct image sheaf $f_o(\mathcal{S})$ is an analytic sheaf over N (a sheaf of ${}_N\mathcal{O}$ -modules); for it is evident from the definition that $f_o(\mathcal{S})$ is a sheaf of modules over the sheaf of rings $f_o({}_M\mathcal{O})$, and as noted above, $f_o({}_M\mathcal{O}) \supset {}_N\mathcal{O}$. The operation of passing to the direct image of an analytic sheaf is basically more complicated than that of passing to the inverse image; so to establish the properties of the direct image mapping f_o which will be required, further restrictions will be imposed on the mapping f .

Lemma 9. Let $f: M \longrightarrow N$ be a complex analytic mapping between two Riemann surfaces, which exhibits M as an r-sheeted branched covering of N ; and let $\mathcal{S}, \mathcal{S}_1, \mathcal{S}_2, \ldots$ be analytic sheaves over the Riemann surface M .

(i) Then $f_o(\mathcal{S}_1 \oplus \mathcal{S}_2) = f_o(\mathcal{S}_1) \oplus f_o(\mathcal{S}_2)$.

(ii) If $\mathcal{S}_j$ are coherent analytic sheaves over M forming an exact sequence $0 \to \mathcal{S}_1 \to \mathcal{S}_2 \to \mathcal{S}_3 \to 0$, then the following is an exact sequence of analytic sheaves over N .

$$0 \longrightarrow f_o(\mathcal{S}_1) \longrightarrow f_o(\mathcal{S}_2) \longrightarrow f_o(\mathcal{S}_3) \longrightarrow 0 .$$

(iii) The direct image sheaf $f_o({}_M\mathcal{O})$ is a locally free analytic sheaf of rank r over N .

(iv) If $\mathcal{S}$ is a coherent analytic sheaf over M , the direct image $f_o(\mathcal{S})$ is a coherent analytic sheaf over N .

Proof. (i) This assertion is obvious, and actually holds for arbitrary complex analytic mappings f .

(ii) For any point $p \in N$ there is an open neighborhood U of p sufficiently small that each connected component V_i of the set $f^{-1}(U) \subset M$ is contained in a coordinate neighborhood, and that the sheaves $\mathcal{S}_j | V_i$ are cokernels of injections of free sheaves over V_i ; thus, by the Corollary to Theorem 3, it follows that $H^1(V_i, \mathcal{S}_j) = 0$. For each set V_i , the exact cohomology sequence associated to the exact sequence of sheaves $0 \to \mathcal{S}_1|V_i \to \mathcal{S}_2|V_i \to \mathcal{S}_3|V_i \to 0$ therefore begins as follows:

$$0 \longrightarrow \Gamma(V_i, \mathcal{S}_1) \longrightarrow \Gamma(V_i, \mathcal{S}_2) \longrightarrow \Gamma(V_i, \mathcal{S}_3) \longrightarrow 0 .$$

Taking the direct sum of these sequences over all the neighborhoods V_i , the following is also an exact sequence.

$$0 \longrightarrow \Gamma(f^{-1}(U), \mathcal{S}_1) \longrightarrow \Gamma(f^{-1}(U), \mathcal{S}_2) \longrightarrow \Gamma(f^{-1}(U), \mathcal{S}_3) \longrightarrow 0 .$$

Taking the direct limit of these exact sequences in turn, leads immediately to the desired result.

(iii) If the point $p \in N$ is not the image of a branch point under the mapping f , then for any sufficiently small connected open neighborhood U of p the inverse image $f^{-1}(U)$ will decompose into the disjoint union of r connected sets V_j , each of which is analytically equivalent to U ; and thus $\Gamma(f^{-1}(U), {}_M\mathcal{O}) = \sum_{j=1}^{r} \Gamma(V_j, {}_M\mathcal{O}) \cong \sum_{j=1}^{r} \Gamma(U, {}_N\mathcal{O})$, considered as $\Gamma(U, {}_N\mathcal{O})$-modules. In a neighborhood of such a point, it follows that $f_o({}_M\mathcal{O}) \cong {}_N\mathcal{O}^r$. If p is the image of a branch point, there is still a decomposition $\Gamma(f^{-1}(U), {}_M\mathcal{O}) = \sum_j \Gamma(V_j, {}_M\mathcal{O})$; but now the number of components V_j is less than r , and some of them are branched coverings at p . It suffices then to consider a single component V_j , and prove the desired result in that case. Suppose therefore that $f: M \longrightarrow N$ exhibits M as an r-sheeted branched covering of N , with a branch point of order $r-1$ over p ; restricting to a sufficiently small neighborhood of p and choosing coordinates properly, it can be assumed that $M = \{z \in \mathbb{C} \mid |z| < 1\}$, $N = \{w \in \mathbb{C} \mid |w| < 1\}$, and $w = f(z) = z^r$. Now if U is any open disc around the point $p: z = 0$, and if $h \in \Gamma(f^{-1}(U), {}_M\mathcal{O})$, then

$$h(z) = \sum_{\nu=0}^{\infty} a_\nu z^\nu = \sum_{\nu=0}^{\infty} (a_{\nu r} z^{\nu r} + z\cdot a_{\nu r+1} z^{\nu r} + \ldots + z^{r-1}\cdot a_{\nu r+r-1} z^{\nu r})$$

$$= h_1(w) + z\cdot h_2(w) + \ldots + z^{r-1} h_{r-1}(w) .$$

Thus $\Gamma(f^{-1}(U), {}_M\mathcal{O}) \cong \Gamma(U, {}_N\mathcal{O})^r$ as $\Gamma(U, {}_N\ )$-modules, the isomorphism being that which associates to $h(z)$ the r-tuple $(h_1(w), \ldots, h_r(w))$. This isomorphism is compatible with the isomorphisms introduced above at all the regular points $z \neq 0$, and therefore $f_o({}_M\mathcal{O}) \cong {}_N\mathcal{O}^r$, as desired.

(iv) Finally, suppose that $\mathcal{S}$ is a coherent analytic sheaf over M. For each point $p \in N$ and each sufficiently small open neighborhood U of p, there is then an exact sequence of analytic sheaves of the following form.

$$0 \longrightarrow {}_M\mathcal{O}^{m_1}|f^{-1}(U) \longrightarrow {}_M\mathcal{O}^{m}|f^{-1}(U) \longrightarrow \mathcal{S}\,|f^{-1}(U) \longrightarrow 0 .$$

Applying (ii), there results the following exact sequence of analytic sheaves over U:

$$0 \longrightarrow f_o({}_M\mathcal{O}^{m_1}) \longrightarrow f_o({}_M\mathcal{O}^{m}) \longrightarrow f_o(\mathcal{S}) \longrightarrow 0 .$$

By (i), $f_o({}_M\mathcal{O}^m) \cong f_o({}_M\mathcal{O})^m$; and this sheaf is then coherent, by (iii). Thus $f_o(\mathcal{S})$ is coherent, and the proof is thereby concluded.

Remark. The analogous result in several complex variables is considerably more difficult, but of great importance; see for instance H. Grauert and R. Remmert, Bilder und Urbilder Analytischer Garben, (Annals of Math. 68(1958), 393-443).

(c) The operations of passing to inverse images and to direct images of sheaves are of course not inverse to one another; there are several interesting relations between the two operations, but for the present purposes, only the following observations need be made.

Lemma 10. Let $f: M \longrightarrow N$ be a complex analytic mapping between two Riemann surfaces, which exhibits M as an r-sheeted branched covering of N; and let $\mathcal{S}$ be an analytic sheaf over M.

(i) There is a homomorphism of analytic sheaves $\varphi: f^o(f_o(\mathcal{S})) \longrightarrow \mathcal{S}$, which is onto.

(ii) If $\mathcal{F}$ is a locally free sheaf of rank 1 over N, then $f_o(f^o(\mathcal{F}) \otimes_{{}_M\mathcal{O}} \mathcal{S}) = \mathcal{F} \otimes_{{}_N\mathcal{O}} f_o(\mathcal{S})$.

Proof. (i) To begin, there is a homomorphism $\varphi': f^{-1}(f_o(\mathcal{S})) \longrightarrow \mathcal{S}$ of sheaves of $f^{-1}({}_N\mathcal{O})$-modules, defined as follows. If $h \in f^{-1}(f_o(\mathcal{S}))_p$ for some point $p \in M$, then by definition, $h \in (f_o(\mathcal{S}))_{f(p)}$. The germ h will have a representative $h_\alpha \in \Gamma(f^{-1}(V_\alpha), \mathcal{S})$ for some open neighborhood V_α of $f(p)$, by the definition of the direct image sheaf. Since $p \in f^{-1}(V_\alpha)$, the section h_α determines an element $h_\alpha(p) \in \mathcal{S}_p$; and this element is defined to be the image $\varphi'(h)$. It is evident that φ' is a well-defined homomorphism of sheaves of $f^{-1}({}_N\mathcal{O})$-modules. Note also that the mapping φ' is onto. For if $s \in \mathcal{S}_p$, and $s_\alpha \in \Gamma(U_\alpha, \mathcal{S})$ is a section of $\mathcal{S}$ passing through s, then s_α can be extended to a section $s_\alpha \in \Gamma(f^{-1}(f(U_\alpha)), \mathcal{S})$, by setting $s_\alpha \equiv 0$ on the components of

$f^{-1}(f(U_\alpha))$ other than U_α. The section s_α determines an element $h \in (f_o(\mathcal{S}))_{f(p)} = f^{-1}(f_o(\mathcal{S}))_p$, and clearly $\varphi'(h) = s$.

Now recalling that $f^o(f_o(\mathcal{S})) = {}_M\mathcal{O} \otimes_{f^{-1}({}_N\mathcal{O})} f^{-1}(f_o(\mathcal{S}))$, define the homomorphism $\varphi\colon f^o(f_o(\mathcal{S})) \longrightarrow \mathcal{S}$ by putting $\varphi(g \otimes h) = g\cdot\varphi'(h)$ whenever $g \in {}_M\mathcal{O}_p$ and $h \in f^{-1}(f_o(\mathcal{S}))_p$; it follows readily that φ is a well-defined homomorphism of analytic sheaves, and φ is onto since φ' is onto.

(ii) After restricting attention to a sufficiently small open neighborhood on N, it can be assumed that $\mathcal{F} = {}_N\mathcal{O}$. Then, on the one hand, $\mathcal{F} \otimes_{{}_N\mathcal{O}} f_o(\mathcal{S}) = {}_N\mathcal{O} \otimes_{{}_N\mathcal{O}} f_o(\mathcal{S}) = f_o(\mathcal{S})$; and on the other hand, $f^o(\mathcal{F}) = f^o({}_N\mathcal{O}) = {}_M\mathcal{O}$ so that $f^o(\mathcal{F}) \otimes_{{}_M\mathcal{O}} \mathcal{S} = {}_M\mathcal{O} \otimes_{{}_M\mathcal{O}} \mathcal{S} = \mathcal{S}$ and $f_o(f^o(\mathcal{F}) \otimes_{{}_M\mathcal{O}} \mathcal{S}) = f_o(\mathcal{S})$ also. This suffices to conclude the proof.

By using the various properties of induced sheaves which have been established above, it is quite easy to extend to arbitrary compact Riemann surfaces the basic existence and representation theorems established in §2 for the projective line. Recall from §10 of last year's lectures that any compact Riemann surface M can be represented as an r-sheeted branched covering $f\colon M \to \mathbb{P}$ of the projective line.

<u>Theorem 8.</u> If $\mathcal{S}$ is a coherent analytic sheaf over a compact Riemann surface M, then there are locally free analytic sheaves $\mathcal{F}$, $\mathcal{F}_1$ over M for which there is an exact sequence of the form

$$0 \longrightarrow \mathcal{F}_1 \longrightarrow \mathcal{F} \longrightarrow \mathcal{S} \longrightarrow 0 .$$

Proof. Let $f: M \longrightarrow \mathbb{P}$ be a complex analytic mapping exhibiting M as an r-sheeted branched covering of the projective line $\mathbb{P}$. The direct image sheaf $f_o(\mathcal{S})$ is a coherent analytic sheaf over $\mathbb{P}$, by Lemma 9(iv); so by Theorem 6, $f_o(\mathcal{S})$ will be the image of a homomorphism $\mathcal{R} \longrightarrow f_o(\mathcal{S})$, for some locally free analytic sheaf $\mathcal{R}$ over the projective line. That is, there is an exact sequence of analytic sheaves over $\mathbb{P}$ of the form $\mathcal{R} \longrightarrow f_o(\mathcal{S}) \longrightarrow 0$. The sheaf $\mathcal{F} = f^o(\mathcal{R})$ is a locally free sheaf by Lemma 8(iii); and by Lemma 8(ii), the following is also an exact sequence of analytic sheaves: $\mathcal{F} \xrightarrow{\varphi} f^o(f_o(\mathcal{S})) \longrightarrow 0$. Combining the homomorphism φ with the homomorphism of Lemma 10(i) leads finally to the exact sequence $\mathcal{F} \xrightarrow{\psi} \mathcal{S} \longrightarrow 0$. The kernel of ψ is coherent by Theorem 2; and as a coherent subsheaf of a locally free sheaf, it must also be locally free, by Theorem 3. This therefore provides the desired exact sequence $0 \longrightarrow \mathcal{F}_1 \longrightarrow \mathcal{F} \longrightarrow \mathcal{S} \longrightarrow 0$.

<u>Theorem 9.</u> If $\mathcal{S}$ is a coherent analytic sheaf over a compact Riemann surface M, there is a complex line bundle ξ over M such that $\mathcal{O}(\xi) \otimes \mathcal{S}$ admits a non-trivial global section.

Proof. Let $f: M \longrightarrow \mathbb{P}$ be a complex analytic mapping exhibiting M as an r-sheeted branched covering of the projective line $\mathbb{P}$. The direct image sheaf $f_o(\mathcal{S})$ is coherent by Lemma 9(iv), so upon recalling the remarks after Theorem 7, it follows that ${}_{\mathbb{P}}\mathcal{M} \otimes_{{}_{\mathbb{P}}\mathcal{O}} f_o(\mathcal{S}) \cong {}_{\mathbb{P}}\mathcal{M}^m$, (where ${}_{\mathbb{P}}\mathcal{M}$ is the sheaf of germs of meromorphic functions over $\mathbb{P}$); there is therefore a non-trivial section g of this tensor product sheaf. For a suitable finite open covering $\{U_\alpha\}$ of $\mathbb{P}$, this section has a representation $g = \{m_\alpha \otimes s_\alpha\}$, where m_α is a meromorphic function in U_α and $s_\alpha \in \Gamma(U_\alpha, f_o(\mathcal{S}))$. The functions m_α will have altogether finitely many poles over $\mathbb{P}$; so there is a locally free sheaf $\mathcal{F}$ of rank 1 over $\mathbb{P}$ with a section $h = \{h_\alpha\} \in \Gamma(\mathbb{P}, \mathcal{F})$ such that $h_\alpha m_\alpha$ is holomorphic in U_α. Then $hg = \{h_\alpha m_\alpha \otimes s_\alpha\}$ is a non-trivial holomorphic section of $\mathcal{F} \otimes f_o(\mathcal{S})$, so that $\Gamma(\mathbb{P}, \mathcal{F} \otimes f_o(\mathcal{S})) \neq 0$. By Lemma 8(iii) the sheaf $f^o(\mathcal{F})$ will be locally free of rank 1 over M, so that $f^o(\mathcal{F}) = \mathcal{O}(\xi)$ for some complex line bundle ξ over M. Then from formula (2) and Lemma 10(ii) it follows that $\Gamma(M, \mathcal{O}(\xi) \otimes \mathcal{S}) = \Gamma(M, f^o(\mathcal{F}) \otimes_{{}_M\mathcal{O}} \mathcal{S}) \cong \Gamma(\mathbb{P}, f_o(f^o(\mathcal{F}) \otimes_{{}_M\mathcal{O}} \mathcal{S})) \cong \Gamma(\mathbb{P}, \mathcal{F} \otimes_{{}_{\mathbb{P}}\mathcal{O}} f_o(\mathcal{S})) \neq 0$, which suffices to conclude the proof.

<u>Corollary.</u> Any complex analytic vector bundle over a compact Riemann surface M admits non-trivial meromorphic sections.

Proof. If $\Phi \in H^1(M, \mathcal{GL}(m, \mathcal{O}))$ is a vector bundle of rank m over M, it follows from Theorem 9 that there is a complex line bundle $\xi \in H^1(M, \mathcal{O}^*)$ such that $\mathcal{O}(\xi) \otimes \mathcal{O}(\Phi) = \mathcal{O}(\xi \otimes \Phi)$ admits a non-trivial holomorphic section h; letting g be a non-trivial meromorphic section of ξ^{-1}, it follows that gh is a non-trivial meromorphic section of Φ, as desired.

§4. Riemann-Roch theorem

(a) Before considering the Riemann-Roch theorem itself, it is necessary to examine some further general properties of complex analytic vector bundles. Let $\Phi \in H^1(M, \mathcal{GL}(m, \mathcal{O}))$ be a complex vector bundle of rank m over a Riemann surface M; and let $(\Phi_{\alpha\beta}) \in Z^1(\mathcal{U}, \mathcal{GL}(m, \mathcal{O}))$ be a representative cocycle, in terms of some open covering $\mathcal{U} = \{U_\alpha\}$ of M. Since the elements $\Phi_{\alpha\beta} \in GL(m, \mathcal{O}_{U_\alpha \cap U_\beta})$ are matrices over a commutative ring, their determinants $\varphi_{\alpha\beta} = \det \Phi_{\alpha\beta} \in GL(1, \mathcal{O}_{U_\alpha \cap U_\beta}) = \mathcal{O}^*_{U_\alpha \cap U_\beta}$ are well-defined; and it is evident that these determinants form a cocycle $(\varphi_{\alpha\beta}) \in Z^1(\mathcal{U}, \mathcal{O}^*)$. The cohomology class φ of the cocycle is clearly determined uniquely by the cohomology class Φ, and is thus a line bundle canonically associated to the vector bundle Φ. The line bundle φ will be called the <u>determinant line bundle</u> of the vector bundle Φ, and will be denoted by $\varphi = \det \Phi$. The elementary properties of the determinant line bundle follow immediately from the definition. Thus for two vector bundles $\Phi_1 \in H^1(M, \mathcal{GL}(m_1, \mathcal{O}))$ and $\Phi_2 \in H^1(M, \mathcal{GL}(m_2, \mathcal{O}))$, note that

$$(1) \quad \begin{cases} \det(\Phi_1 \oplus \Phi_2) = (\det \Phi_1)\cdot(\det \Phi_2), \text{ and} \\ \det(\Phi_1 \otimes \Phi_2) = (\det \Phi_1)^{m_2}\cdot(\det \Phi_2)^{m_1}, \end{cases}$$

where the tensor product of line bundles (the group operation in $H^1(M, \mathcal{O}^*)$) is written as multiplication. In particular, if ξ is a complex line bundle and Φ is a complex vector bundle of rank m, then $\det(\xi \otimes \Phi) = \xi^m \cdot \det \Phi$. Further, if $\Phi \subset \Psi$ and $\Psi/\Phi = \Theta$, then $\det \Psi = (\det \Phi) \cdot (\det \Theta)$. The Chern class, or characteristic class, of a complex line bundle ξ , considered as an element of the group $\mathbb{Z}$ of the integers, will be denoted by $c(\xi)$. (For the general properties and the definition of the Chern class of a line bundle, recall §7(a) of last year's lectures.) Thus to a complex vector bundle Φ is associated the integer $c(\det \Phi)$, the Chern class of its determinant line bundle. (This integer is sometimes called the degree of the line bundle, but that terminology will not be used in the present lectures.)

A complex vector bundle $\Phi \in H^1(M, \mathcal{GL}(m, \mathcal{O}))$ is called a subbundle of a complex vector bundle $\Psi \in H^1(M, \mathcal{GL}(n, \mathcal{O}))$, and written $\Phi \subset \Psi$, if the corresponding locally free sheaf $\mathcal{O}(\Phi)$ is given as a subsheaf of $\mathcal{O}(\Psi)$, in such a manner that the quotient sheaf $\mathcal{O}(\Psi)/\mathcal{O}(\Phi)$ is also locally free. The quotient sheaf is then associated to a complex vector bundle $\Theta \in H^1(M, \mathcal{GL}(r, \mathcal{O}))$, which will be called the quotient bundle and denoted by $\Theta = \Psi/\Phi$. The bundle Ψ is also called an extension of the bundle Φ by the bundle Θ . In such a situation, there is an exact sequence of locally free analytic sheaves of the form

$$0 \longrightarrow \mathcal{O}(\Phi) \longrightarrow \mathcal{O}(\Psi) \longrightarrow \mathcal{O}(\Theta) \longrightarrow 0 ;$$

the associated exact cohomology sequence then begins

$$0 \longrightarrow \Gamma(M, \mathcal{O}(\Phi)) \longrightarrow \Gamma(M, \mathcal{O}(\Psi)) \longrightarrow \Gamma(M, \mathcal{O}(\Theta)) \longrightarrow \dots ,$$

so that there are natural mappings from the space of holomorphic sections of a subbundle $\Phi \subset \Psi$ to the space of holomorphic sections of the ambient bundle Ψ , and from the space of holomorphic sections of a bundle Ψ to the space of holomorphic sections of a quotient bundle Ψ/Φ . A vector bundle Ψ is called reducible if it has a proper subbundle; otherwise Ψ is called irreducible. A vector bundle Ψ is said to be decomposable into the direct sum of two vector bundles Φ and Θ , written $\Psi = \Phi \oplus \Theta$, if the corresponding locally free sheaf $\mathcal{O}(\Psi)$ is given as the direct sum $\mathcal{O}(\Psi) = \mathcal{O}(\Phi) \oplus \mathcal{O}(\Theta)$; if no such decomposition is possible, the bundle Ψ is said to be indecomposable. If the vector bundle Ψ is decomposable as $\Psi = \Phi \oplus \Theta$, then both Φ and Θ are subbundles, so that Ψ is certainly reducible; or what is the same thing, an irreducible bundle is indecomposable. One of the complicating factors in the study of complex vector bundles is that the converse assertion is generally false; that is, an indecomposable bundle need not be irreducible.

To examine these latter definitions a bit further, let $(\Phi_{\alpha\beta}) \in Z^1(\mathcal{U}, \mathcal{GL}(m, \mathcal{O}))$ and $(\Psi_{\alpha\beta}) \in Z^1(\mathcal{U}, \mathcal{GL}(n, \mathcal{O}))$ be defining cocycles for complex vector bundles Φ and Ψ over a Riemann surface M ; and suppose that $\Phi \subset \Psi$ and $\Psi/\Phi = \Theta$. Over each coordinate neighborhood U_α there are then isomorphisms $\mathcal{O}(\Phi)|U_\alpha \cong \mathcal{O}^m|U_\alpha$ and $\mathcal{O}(\Psi)|U_\alpha \cong \mathcal{O}^n|U_\alpha$; and the imbedding $\mathcal{O}^m|U_\alpha \subset \mathcal{O}^n|U_\alpha$ is such that the quotient is a free bundle of rank r over U_α . It is evident that there is no loss of generality in putting $\mathcal{O}^n|U_\alpha = \mathcal{O}^m|U_\alpha \oplus \mathcal{O}^r|U_\alpha$. (For let E_j be the canonical generating sections of $\mathcal{O}^r|U_\alpha$, and, passing to a refinement if necessary, let F_j be sections of $\mathcal{O}^n|U_\alpha$ which are the images of E_j under the mapping to the quotient sheaf. There is an injection $\mathcal{O}^r|U_\alpha \longrightarrow \mathcal{O}^n|U_\alpha$ defined by taking $F_j \longrightarrow E_j$; and it is easily seen that $\mathcal{O}^n|U_\alpha$ is the direct sum of $\mathcal{O}^m|U_\alpha$ and the image of $\mathcal{O}^r|U_\alpha$.) Thus an element $F \in \mathcal{O}^m_p$ corresponds to an n-tuple $\binom{F}{0} \in \mathcal{O}^n_p$ under this imbedding. The matrices $\Psi_{\alpha\beta}$ must preserve the subspaces $\mathcal{O}^m_p \subset \mathcal{O}^n_p$; indeed, their restrictions to the subspaces $\mathcal{O}^m_p$ must be the matrices $\Phi_{\alpha\beta}$, and their effect on the quotient spaces must be given by the matrices $\Theta_{\alpha\beta}$. Therefore the matrices $\Psi_{\alpha\beta}$ must necessarily have the form

$$(2) \qquad \Psi_{\alpha\beta} = \begin{pmatrix} \Phi_{\alpha\beta} & \Lambda_{\alpha\beta} \\ 0 & \Theta_{\alpha\beta} \end{pmatrix}$$

The converse being apparent, the condition that $\Phi \subset \Psi$ and $\Psi/\Phi = \Theta$ is therefore just that the bundles can be represented by cocycles such that (2) holds. The condition that Ψ be decomposable into the direct sum $\Psi = \Phi \oplus \Theta$ is correspondingly that the bundles can be represented by cocycles of the form (2) with $\Lambda_{\alpha\beta} = 0$.

<u>Lemma 11.</u> Let Ψ be a complex vector bundle of rank $m > 1$ over a Riemann surface M, and suppose that Ψ admits a non-trivial meromorphic section F. Then Ψ has as subbundle a complex line bundle ψ; and ψ has a meromorphic section f which determines the section F under the natural injection

$$0 \longrightarrow \mathcal{M}(\psi) \longrightarrow \mathcal{M}(\Psi) .$$

(That is, $f = F$ when f is considered as a section of the larger bundle.)

Proof. Let $\Psi_{\alpha\beta} \in Z^1(\mathfrak{U}, \mathcal{GL}(m, \mathcal{O}))$ be a defining cocycle for the complex vector bundle Ψ, in terms of some coordinate covering $\mathfrak{U} = \{U_\alpha\}$ of M; and let $F = \{F_\alpha\}$ be a non-trivial meromorphic section of this bundle. Thus $F_\alpha \in \Gamma(U_\alpha, \mathcal{M}^m)$ is not identically zero, and $F_\alpha(p) = \Psi_{\alpha\beta}(p)\cdot F_\beta(p)$ whenever $p \in U_\alpha \cap U_\beta$; here F_α is envisaged as a column vector of meromorphic functions. Refining the covering $\mathfrak{U}$ if necessary, suppose that F_α is holomorphic and non-singular (that is, not all components vanish) at all points in U_α except perhaps one point; and let z_α be a coordinate mapping in U_α such that the exceptional point is the origin $z_\alpha = 0$. Then for some integer r_α, $z_\alpha^{r_\alpha} F_\alpha(z_\alpha)$ is a holomorphic, non-singular vector function on all of U_α. Again refining $\mathfrak{U}$ if necessary, there is a holomorphic non-singular matrix valued function $\Psi_\alpha \in GL(m, \mathcal{O}_{U_\alpha})$ such that

$$\Psi_\alpha \cdot z_\alpha^{r_\alpha} F_\alpha = E_1 ,$$

where as before E_1 is the constant vector

$$E_1 = \begin{pmatrix} 1 \\ 0 \\ \cdots \\ 0 \end{pmatrix} \in \mathbf{C}^m .$$

The vector bundle Ψ is also defined by the equivalent cocycle $\Psi'_{\alpha\beta} = \Psi_\alpha \Psi_{\alpha\beta} \Psi_\beta^{-1}$, and the meromorphic section F is expressed in terms of this cocycle by the functions $F'_\alpha = \Psi_\alpha F_\alpha = z_\alpha^{-r_\alpha} E_1$. Since $z_\alpha^{-r_\alpha} E_1 = \Psi'_{\alpha\beta} z_\beta^{-r_\beta} E_1$ in $U_\alpha \cap U_\beta$, it is apparent that the matrix $\Psi'_{\alpha\beta}$ must be of the form

$$\Psi'_{\alpha\beta} = \begin{pmatrix} \psi_{\alpha\beta} & * & \dots & * \\ 0 & * & \dots & * \\ \dots & & & \dots \\ 0 & * & \dots & * \end{pmatrix} ;$$

here $(\psi_{\alpha\beta})$ defines a line bundle $\psi \subset \Psi$ with a meromorphic section $(f_\alpha) = (z_\alpha^{-r_\alpha})$ inducing the given section $F = (F'_\alpha)$, and the proof is thereby concluded.

Theorem 10. Every complex vector bundle of rank $m > 1$ over a compact Riemann surface contains a line bundle as a subbundle; so only line bundles are irreducible.

Proof. The Corollary to Theorem 9 shows that every complex vector bundle over a compact Riemann surface admits a nontrivial meromorphic section; so the theorem follows as an immediate consequence of Lemma 11.

Corollary. A complex vector bundle $\Psi \in H^1(M, \mathfrak{b}\mathfrak{L}(m, \mathcal{O}))$ over a compact Riemann surface M can always be defined by a cocycle $(\psi_{\alpha\beta}) \in Z^1(\mathfrak{U}, \mathfrak{b}\mathfrak{L}(m, \mathcal{O}))$ of the form

$$(3) \qquad \Psi_{\alpha\beta} = \begin{pmatrix} \psi_{1\alpha\beta} & * & * & \dots & * \\ 0 & \psi_{2\alpha\beta} & * & \dots & * \\ 0 & 0 & \psi_{3\alpha\beta} & \dots & * \\ \dots & \dots & \dots & & \dots \\ 0 & 0 & 0 & \dots & \psi_{m\alpha\beta} \end{pmatrix}$$

Proof. The vector bundle Ψ must contain a line bundle ψ_1 as a subbundle, by Theorem 10; so the defining cocycle can be given by matrices having as first column the vector

$$\begin{pmatrix} \psi_{1\alpha\beta} \\ 0 \\ 0 \\ \cdots \\ 0 \end{pmatrix},$$

where $(\psi_{1\alpha\beta})$ defines the line bundle ψ_1. The $(m-1) \times (m-1)$ matrix block along the diagonal (omitting the first row and column) defines the quotient bundle, and then must have the indicated form by the obvious induction.

(b) The Riemann-Roch theorem, for complex line bundles over a compact Riemann surface, was proved in last year's lectures, (Theorem 13, page 111); recall that the theorem can be stated in the following form. If M is a compact Riemann surface of genus g, and $\xi \in H^1(M, \mathcal{O}^*)$ is a complex line bundle of Chern class $c(\xi) \in \mathbb{Z}$, then

$$\dim H^0(M, \mathcal{O}(\xi)) - \dim H^1(M, \mathcal{O}(\xi)) = c(\xi) + 1 - g .$$

This theorem extends easily to complex vector bundles, in the following form.

<u>Theorem 11.</u> (Riemann-Roch theorem for vector bundles). If M is a compact Riemann surface of genus g and $\Phi \in H^1(M, \mathcal{GL}(m, \mathcal{O}))$ is a complex analytic vector bundle of rank m over M, then

$$\dim H^0(M, \mathcal{O}(\Phi)) - \dim H^1(M, \mathcal{O}(\Phi)) = c(\det \Phi) + m(1-g) ,$$

the cohomology groups being finite-dimensional complex vector spaces.

Proof. The proof will be by induction on the rank m. The case $m = 1$ is just the Riemann-Roch Theorem for complex line bundles, and is known. Assuming that the theorem has also been proved for complex vector bundles of rank $m-1$, consider a vector bundle $\Phi \in H^1(M, \mathcal{GL}(m, \mathcal{O}))$ of rank m. By Theorem 10, the bundle Φ contains a line bundle φ as subbundle; so the quotient bundle $\Theta = \Phi/\varphi$ is a well-defined vector bundle of rank $m-1$. To the exact sequence of sheaves

$$0 \longrightarrow \mathcal{O}(\varphi) \longrightarrow \mathcal{O}(\Phi) \longrightarrow \mathcal{O}(\Theta) \longrightarrow 0$$

there corresponds an exact sequence of cohomology groups, beginning as follows.

$$\begin{aligned} &0 \longrightarrow H^0(M, \mathcal{O}(\varphi)) \longrightarrow H^0(M, \mathcal{O}(\Phi)) \longrightarrow H^0(M, \mathcal{O}(\Theta)) \longrightarrow \\ (4) \qquad &\longrightarrow H^1(M, \mathcal{O}(\varphi)) \longrightarrow H^1(M, \mathcal{O}(\Phi)) \longrightarrow H^1(M, \mathcal{O}(\Theta)) \longrightarrow \\ &\longrightarrow H^2(M, \mathcal{O}(\varphi)) \longrightarrow \cdots \end{aligned}$$

It follows immediately from this exact sequence and the induction hypothesis that all the cohomology groups are finite-dimensional complex vector spaces; and, recalling that $H^2(M, \mathcal{O}(\varphi)) = 0$ for any line bundle φ, the alternating sum of the dimensions of the complex vector spaces in the exact sequence (4), up to the term $H^2(M, \mathcal{O}(\varphi))$, must be zero. Writing that sum out, and regrouping terms, it follows from the induction hypothesis that

$$\dim H^0(M, \mathcal{O}(\Phi)) - \dim H^1(M, \mathcal{O}(\Phi))$$
$$= [\dim H^0(M, \mathcal{O}(\varphi)) - \dim H^1(M, \mathcal{O}(\varphi))] +$$
$$+ [\dim H^0(M, \mathcal{O}(\Theta)) - \dim H^1(M, \mathcal{O}(\Theta))]$$
$$= [c(\varphi) + 1 - g] + [c(\det \Theta) + (m-1)(1-g)] ;$$

but since $\det \Phi = (\det \varphi)\cdot(\det \Theta)$, and hence $c(\det \Phi) = c(\varphi) + c(\det \Theta)$, the induction is completed and the proof therewith concluded.

Corollary. For any complex vector bundle $\Phi \in H^1(M, \mathcal{GL}(m, \mathcal{O}))$ over a compact Riemann surface M ,

$$H^q(M, \mathcal{O}(\Phi)) = 0 \quad \text{for all} \quad q > 1 .$$

Proof. The proof is again by induction on the rank m of Φ , the case $m = 1$ having been demonstrated in last year's lectures. With the notation as in the proof of Theorem 11, the exact sequence (4) contains segments of the form

$$\longrightarrow H^q(M, \mathcal{O}(\varphi)) \longrightarrow H^q(M, \mathcal{O}(\Phi)) \longrightarrow H^q(M, \mathcal{O}(\Theta)) \longrightarrow \cdots$$

for all $q > 1$; but since $H^q(M, \mathcal{O}(\varphi)) = 0$ and $H^q(M, \mathcal{O}(\Theta)) = 0$ by the induction hypothesis, the desired result follows immediately.

Remarks. For any coherent analytic sheaf $\mathcal{S}$ on a compact Riemann surface M , it follows from Theorem 8 that there are complex vector bundles Φ, Φ_1 , forming an exact sequence

$$(5) \qquad 0 \longrightarrow \mathcal{O}(\Phi_1) \longrightarrow \mathcal{O}(\Phi) \longrightarrow \mathcal{S} \longrightarrow 0 .$$

It is an immediate consequence of the above results and the exact cohomology sequence associated to (5) that the groups $H^q(M, \mathcal{S})$ are finite-dimensional complex vector spaces, and that $H^q(M, \mathcal{S}) = 0$ whenever $q > 1$. Writing

$$\chi(\mathcal{S}) = \dim H^0(M, \mathcal{S}) - \dim H^1(M, \mathcal{S}) ,$$

the exact cohomology sequence associated to (5) also shows that

$$\chi(\mathcal{S}) = \chi(\Phi) - \chi(\Phi_1) ;$$

this provides a form of the Riemann-Roch theorem for arbitrary coherent analytic sheaves. (In this connection, see also the article by A. Borel and J.-P. Serre, Le théorème de Riemann-Roch, Bull. Soc. Math., France 86(1958), 97-136.)

(c) The Serre duality theorem also extends readily to complex vector bundles. (The line bundle case was treated in last year's lectures, §4 and §5.) Recall that over any Riemann surface M there is the Dolbeault exact sequence of analytic sheaves, of the form

$$(6) \qquad 0 \longrightarrow \mathcal{O} \longrightarrow \mathcal{E}^{0,0} \xrightarrow{\bar{\partial}} \mathcal{E}^{0,1} \longrightarrow 0 ,$$

where $\mathcal{E}^{r,s}$ is the sheaf of germs of C^∞ differential forms of type (r,s). (See page 72 of last year's lectures; that these sheaves $\mathcal{E}^{r,s}$ can be viewed as sheaves of $\mathcal{O}$-modules, and that $\bar{\partial}$ is a homomorphism of sheaves of $\mathcal{O}$-modules, are quite evident.) For any complex analytic vector bundle Φ over M, the sheaf $\mathcal{O}(\Phi)$ is locally free, hence flat, so tensoring with (6) yields the exact sequence of analytic sheaves

(7) $$0 \longrightarrow \mathcal{O}(\Phi) \longrightarrow \mathcal{E}^{0,0}(\Phi) \xrightarrow{\bar\partial} \mathcal{E}^{0,1}(\Phi) \longrightarrow 0 ;$$

here $\mathcal{E}^{r,s}(\Phi) = \mathcal{E}^{r,s} \otimes_{\mathcal{O}} \mathcal{O}(\Phi)$ is the sheaf of germs of C^∞ differential forms of type (r,s) which are sections of the vector bundle Φ. (See page 73 of last year's lectures.) It is clear that $\mathcal{E}^{r,s}(\Phi)$ is a fine sheaf; and so, recalling Theorem 3 of last year's lectures,

(8) $$H^1(M, \mathcal{O}(\Phi)) \cong \frac{\Gamma(M, \mathcal{E}^{0,1}(\Phi))}{\bar\partial\Gamma(M, \mathcal{E}^{0,0}(\Phi))} .$$

Letting $\mathcal{K}$ be the sheaf of germs of distributions on the Riemann surface M, (recalling §5 of last year's lectures), there is an exact sequence parallel to (7), replacing the C^∞ differential forms by distribution differential forms; and there follows an analogue of formula (8) for distributions.

The spaces $\Gamma(M, \mathcal{E}^{1,1}(\Phi))$ can be given the structures of topological vector spaces, as follows. Let $\{U_\alpha, z_\alpha\}$ be a coordinate covering of the compact surface M by a finite number of coordinate neighborhoods which are small enough that $\mathcal{O}(\Phi)$ is free over each U_α. A section $F \in \Gamma(M, \mathcal{E}^{1,1}(\Phi))$ can then be written $F = \{F_\alpha\}$, where $F_\alpha \in \Gamma(U_\alpha, (\mathcal{E}^{1,1})^m)$ and m is the rank of the vector bundle Φ; thus $F_\alpha = (f_{i\alpha} dz^\alpha \wedge d\bar z_\alpha)$, where $f_{i\alpha} \in \Gamma(U_\alpha, \mathcal{E}^{0,0})$. For any integer $n \geqq 0$, put

$$p_n(F) = \sum_\alpha \sum_i \sum_{\nu_1+\nu_2 \leq n} \sup_{z_\alpha \in V_\alpha} |D^\nu f_{i\alpha}(z_\alpha)| ,$$

with the notation as in §5 of last year's lectures. The p_n are norms on $\Gamma(M, \mathcal{E}^{1,1}(\Phi))$, and thus determine the structure of a topological vector space. Of course, since $\Gamma(M, \mathcal{E}^{r,s}(\Phi)) =$ $= \Gamma(M, \mathcal{E}^{1,1}(\Phi \otimes \kappa^{1-r} \otimes \kappa^{1-s})$, all the spaces $\Gamma(M, \mathcal{E}^{r,s}(\Phi))$ can likewise be topologized.

Now, for the application of these considerations, let $\Phi \in H^1(M, \mathcal{GL}(m, \mathcal{O}))$ be any complex vector bundle over the compact Riemann surface M, and let $(\Phi_{\alpha\beta}) \in Z^1(\mathfrak{U}; \mathcal{GL}(m, \mathcal{O}))$ be a defining cocycle in terms of a coordinate covering $\mathfrak{U} = \{U_\alpha, z_\alpha\}$ of M. The matrices $\Phi^*_{\alpha\beta} = {}^t\Phi^{-1}_{\alpha\beta} \in GL(m, \mathcal{O}_{U_\alpha \cap U_\beta})$, where t denotes the transposed matrix, clearly also define a cocycle for that covering $\mathfrak{U}$; the bundle defined by this cocycle $(\Phi^*_{\alpha\beta})$ will be called the <u>dual bundle</u> to the complex vector bundle Φ, and will be denoted by Φ^*.

<u>Lemma 12.</u> On a compact Riemann surface M, the space $\Gamma(M, \mathcal{K}(\Phi^*))$ is the space of continuous linear functionals on the topological vector space $\Gamma(M, \mathcal{E}^{1,1}(\Phi))$.

Proof. In terms of a suitable coordinate covering $\mathfrak{U} = \{U_\alpha, z_\alpha\}$, let $T = \{T_\alpha\} \in \Gamma(M, \mathcal{K}(\Phi^*))$ and $F = \{F_\alpha\} \in \Gamma(M, \mathcal{E}^{1,1}(\Phi))$; and let $\{r_\alpha\}$ be a C^∞ partition of unity subordinate to the covering $\mathfrak{U}$. Thus T_α will be an m-tuple of distributions $T_{i\alpha}$ over the open set U_α, and F_α will be an m-tuple of C^∞ differential forms $\theta_{i\alpha}$ of type (1,1) over the open set U_α; and writing the matrices $\Phi_{\alpha\beta} = (\varphi_{ij\alpha\beta})$, $\Phi^*_{\alpha\beta} = {}^t\Phi^{-1}_{\alpha\beta} = (\varphi^*_{ij\alpha\beta})$, in each intersection $U_\alpha \cap U_\beta$ it follows that

$$T_{i\alpha} = \sum_j \varphi^*_{ij\alpha\beta} T_{j\beta} \quad \text{and} \quad \theta_{i\alpha} = \sum_j \varphi_{ij\alpha\beta}\theta_{j\beta} .$$

Each $T_{i\alpha}$ will be a continuous linear functional on the subspace of $\Gamma(M, \mathcal{E}^{1,1})$ consisting of those forms with support in U_α, as noted last year; and hence $T_{i\alpha}(r_\alpha \theta_{i\alpha})$ is well-defined. Note further that, restricting attention to the subset $U_\alpha \cap U_\beta$,

$$\sum_i T_{i\alpha}(r_\alpha r_\beta \theta_{i\alpha}) = \sum_{ijk} \varphi^*_{ij\alpha\beta} T_{j\beta}(r_\alpha r_\beta \varphi_{ik\alpha\beta} \theta_{k\beta}) =$$

$$= \sum_{ijk} \varphi^*_{ij\alpha\beta} \varphi_{ik\alpha\beta} T_{j\beta}(r_\alpha r_\beta \theta_{k\beta}) = \sum_j T_{j\beta}(r_\alpha r_\beta \theta_{j\beta}) .$$

It follows readily that defining

$$T(F) = \sum_\alpha T_\alpha(r_\alpha F_\alpha)$$

exhibits $\Gamma(M, \mathcal{K}(\Phi^*))$ as the space of continuous linear functionals on $\Gamma(M, \mathcal{E}^{1,1}(\Phi))$, as desired.

Theorem 12. (Serre duality for vector bundles.) Let M be a compact Riemann surface, and $\Phi \in H^1(M, \mathcal{GL}(m, \mathcal{O}))$ be a complex vector bundle over M. Then the vector spaces $H^1(M, \mathcal{O}(\Phi))$ and $H^0(M, \mathcal{O}^{1,0}(\Phi^*))$ are canonically dual to one another.

Proof. The proof is the same as that of Theorem 9 of last year's lectures, (replacing Lemma 13 of last year's lectures by the preceding Lemma 12). Thus no further details will be given here.

Remarks. It should be observed that, applying the Serre duality theorem in the form it is given in Theorem 12, the Riemann-Roch theorem for complex vector bundles can be rewritten as

(9) $\quad \dim H^0(M, \mathcal{O}(\Phi)) - \dim H^0(M, \mathcal{O}^{1,0}(\Phi^*)) = c(\det \Phi) + m(1-g)$,

for any vector bundle $\Phi \in H^1(M, \mathcal{GL}(m, \mathcal{O}))$ over a compact Riemann surface M of genus g.

§5. A classification of vector bundles of rank two

(a) In discussing the classification of complex analytic vector bundles over a compact Riemann surface, we shall initially restrict attention to vector bundles of rank two; this case is both sufficiently complicated and sufficiently useful to be of interest. (As to the usefulness, recall the discussion in §9 of last year's lectures.)

Perhaps the first topic that comes to mind in considering the classification problem is that of determining the possible extension of one complex line bundle by another. Let φ_1, φ_2 be two complex analytic line bundles over a Riemann surface M. Recall from §4(a) that a complex analytic vector bundle Φ of rank two is called an <u>extension</u> of φ_1 by φ_2 if there is given an exact sequence of analytic sheaves of the form

$$0 \longrightarrow \mathcal{O}(\varphi_1) \longrightarrow \mathcal{O}(\Phi) \longrightarrow \mathcal{O}(\varphi_2) \longrightarrow 0 \; . \tag{1}$$

Two such extensions Φ and Φ' are called equivalent if there is a sheaf isomorphism $\theta\colon \mathcal{O}(\Phi) \longrightarrow \mathcal{O}(\Phi')$ such that the following diagram is commutative:

$$\begin{array}{ccccccccc} 0 & \longrightarrow & \mathcal{O}(\varphi_1) & \longrightarrow & \mathcal{O}(\Phi) & \longrightarrow & \mathcal{O}(\varphi_2) & \longrightarrow & 0 \\ & & \big\downarrow \text{id} & & \big\downarrow \theta & & \big\downarrow \text{id} & & \\ 0 & \longrightarrow & \mathcal{O}(\varphi_1) & \longrightarrow & \mathcal{O}(\Phi') & \longrightarrow & \mathcal{O}(\varphi_2) & \longrightarrow & 0 \; , \end{array} \tag{2}$$

where <u>id</u> denotes the identity isomorphism; it is evidently only the equivalence classes of extensions that are of importance here.

Theorem 13. If φ_1, φ_2 are complex line bundles over an arbitrary Riemann surface M, the set of equivalence classes of extensions of the line bundle φ_1 by the line bundle φ_2 is in a natural one-to-one correspondence with the set of elements of the cohomology group $H^1(M, \mathcal{O}(\varphi_1\varphi_2^{-1}))$; the trivial extension $\varphi_1 \oplus \varphi_2$ corresponds to the zero element of the group.

Proof. If Φ is an extension of the bundle φ_1 by the bundle φ_2, select a sufficiently fine coordinate covering $\mathfrak{U} = \{U_\alpha\}$ of the Riemann surface M that the bundles involved can be represented by cocycles

$$(\varphi_{j\alpha\beta}) \in Z^1(\mathfrak{U}, \mathcal{O}^*) \quad \text{and} \quad (\Phi_{\alpha\beta}) \in Z^1(\mathfrak{U}, \mathcal{GL}(2, \mathcal{O})).$$

The condition that Φ is an extension of φ_1 by φ_2, that is, that there is an exact sequence of the form (1), is just that the cocycle $(\Phi_{\alpha\beta})$ has the form

$$\Phi_{\alpha\beta} = \begin{pmatrix} \varphi_{1\alpha\beta} & \lambda_{\alpha\beta} \\ 0 & \varphi_{2\alpha\beta} \end{pmatrix}$$

for some elements $\lambda_{\alpha\beta} \in \mathcal{O}_{U_\alpha \cap U_\beta}$. The only restriction upon the elements $(\lambda_{\alpha\beta})$ is that they are so chosen that the matrices $\Phi_{\alpha\beta}$ satisfy the cocycle condition, namely, $\Phi_{\alpha\beta}(p)\Phi_{\beta\gamma}(p) = \Phi_{\alpha\gamma}(p)$ whenever $p \in U_\alpha \cap U_\beta \cap U_\gamma$. Writing this out explicitly, the elements $\lambda_{\alpha\beta}$ must satisfy

$$(3) \quad \lambda_{\alpha\gamma}(p) = \varphi_{1\alpha\beta}(p)\lambda_{\beta\gamma}(p) + \lambda_{\alpha\beta}(p)\varphi_{2\beta\gamma}(p) \quad \text{whenever} \quad p \in U_\alpha \cap U_\beta \cap U_\gamma .$$

It should be noted that the set of functions $(\lambda_{\alpha\beta})$ for a covering $\mathfrak{U}$ induces in a natural manner a corresponding set of functions describing the same extension for any refinement $\mathcal{V}$ of $\mathcal{U}$. Thus, all possible extensions are described by classes of families of functions $(\lambda_{\alpha\beta})$ satisfying (3). Suppose next that Φ and Φ' are two extensions of the same line bundles, defined by sets of functions $(\lambda_{\alpha\beta})$ and $(\lambda'_{\alpha\beta})$ in terms of the same covering of M. Since $\mathcal{O}(\Phi)$ and $\mathcal{O}(\Phi')$ are both free sheaves over the open set U_α, any isomorphism $\theta\colon \mathcal{O}(\Phi) \longrightarrow \mathcal{O}(\Phi')$ is represented over each set U_α by a holomorphic matrix $\Theta_\alpha \in GL(2, \mathcal{O}_{U_\alpha})$; and these matrices clearly satisfy

$$(4) \qquad \Phi'_{\alpha\beta}(p)\cdot\Theta_\beta(p) = \Theta_\alpha(p)\cdot\Phi_{\alpha\beta}(p) \quad \text{whenever} \quad p \in U_\alpha \cap U_\beta .$$

The condition that Φ and Φ' be equivalent extensions is that there exists an isomorphism $\theta\colon \mathcal{O}(\Phi) \longrightarrow \mathcal{O}(\Phi')$ such that (2) is commutative; thus θ must be the identity on the subsheaf $\mathcal{O}(\varphi_1) \subset \mathcal{O}(\Phi)$, and must induce the identity on the quotient sheaf $\mathcal{O}(\varphi_2) = \mathcal{O}(\Phi)/\mathcal{O}(\varphi_1)$. In terms of the matrix representation of the isomorphism θ, this condition is just that the matrices Θ_α be of the form

$$\Theta_\alpha = \begin{pmatrix} 1 & h_\alpha \\ 0 & 1 \end{pmatrix}$$

for some functions $h_\alpha \in \mathcal{O}_{U_\alpha}$. Condition (4) then takes the form

$$(5) \quad \varphi_{1\alpha\beta}(p)h_\beta(p) + \lambda'_{\alpha\beta}(p) = \lambda_{\alpha\beta}(p) + h_\alpha(p)\varphi_{2\alpha\beta}(p) \quad \text{whenever} \quad p \in U_\alpha \cap U_\beta .$$

Thus extensions described by functions $(\lambda_{\alpha\beta})$ and $(\lambda'_{\alpha\beta})$ are equivalent if and only if, possibly after passing to a refinement of the covering, there are holomorphic functions h_α in the various sets U_α such that (5) holds.

Considering in place of the functions $\lambda_{\alpha\beta}$ the functions

$$\sigma_{\alpha\beta} = \varphi_{1\alpha\beta}^{-1}\lambda_{\alpha\beta} ,$$

condition (3) takes the form

(6) $\sigma_{\alpha\gamma}(p) = \sigma_{\beta\gamma}(p) + \sigma_{\alpha\beta}(p)\varphi_{1\beta\gamma}^{-1}(p)\varphi_{2\beta\gamma}(p)$ whenever $p \in U_\alpha \cap U_\beta \cap U_\gamma$.

To the function germ $\sigma_{\alpha\beta}$ at any point p there is canonically associated an element of the sheaf $\mathcal{O}(\varphi_1\varphi_2^{-1})$; in a coordinate neighborhood U_β containing p the element is defined by $\sigma_{\alpha\beta}(z_\beta)$, while if $p \in U_\gamma$ as well, the element is given by $\sigma_{\alpha\beta}(z_\gamma) = \varphi_{1\gamma\beta}\varphi_{2\gamma\beta}^{-1}\sigma_{\alpha\beta}(z_\beta)$ in the coordinate neighborhood U_γ . With this interpretation, (6) takes the form

$$\sigma_{\alpha\gamma}(z_\gamma) = \sigma_{\beta\gamma}(z_\gamma) + \sigma_{\alpha\beta}(z_\beta)\varphi_{1\beta\gamma}^{-1}\varphi_{2\beta\gamma} = \sigma_{\beta\gamma}(z_\gamma) + \sigma_{\alpha\beta}(z_\gamma) ;$$

thus these functions form a cocycle $(\sigma_{\alpha\beta}) \in Z^1(\mathfrak{U}, \mathcal{O}(\varphi_1\varphi_2^{-1}))$. In terms of the functions $\sigma_{\alpha\beta}$, the equivalence condition (5) for two extensions takes the form

(7) $h_\beta(p) + \sigma'_{\alpha\beta}(p) = \sigma_{\alpha\beta}(p) + h_\alpha(p)\varphi_{1\alpha\beta}^{-1}(p)\varphi_{2\alpha\beta}(p)$ whenever $p \in U_\alpha \cap U_\beta$.

Letting $h_\alpha(z_\alpha)$ be the elements of the sheaf $\mathcal{O}(\varphi_1\varphi_2^{-1})$ associated to the functions $h_\alpha(p)$, condition (7) becomes

$$\sigma_{\alpha\beta}(z_\beta) - \sigma'_{\alpha\beta}(z_\beta) = h_\beta(z_\beta) - h_\alpha(z_\alpha)\varphi_{1\alpha\beta}^{-1}(p)\varphi_{2\alpha\beta}(p) = h_\beta(z_\beta) - h_\alpha(z_\beta) ;$$

thus the extensions are equivalent precisely when their defining cocycles are cohomologous, which suffices to conclude the proof. Note that the zero element of the cohomology group obviously corresponds to the trivial extension $\Phi = \varphi_1 \oplus \varphi_2$ as a direct sum.

One immediate consequence of this description of extensions is the following result.

Corollary. Let M be a compact Riemann surface of genus g and Φ be a complex analytic vector bundle of rank two over M. If there is a line bundle $\varphi_1 \subset \Phi$ with quotient bundle $\varphi_2 = \Phi/\varphi_1$, and if $c(\varphi_1) - c(\varphi_2) > 2g-2$, then Φ is decomposable into the direct sum

$$\Phi = \varphi_1 \oplus \varphi_2 .$$

Proof. The set of all extensions of the line bundle φ_1 by the line bundle φ_2 correspond to the elements of the group $H^1(M, \mathcal{O}(\varphi_1\varphi_2^{-1}))$; and by the Serre duality theorem for complex line bundles,

$$H^1(M, \mathcal{O}(\varphi_1\varphi_2^{-1})) \cong H^0(M, \mathcal{O}(\kappa\varphi_1^{-1}\varphi_2)) .$$

If $c(\varphi_1) - c(\varphi_2) > 2g-2$, then $c(\kappa\varphi_1^{-1}\varphi_2) = c(\kappa) - c(\varphi_1) + c(\varphi_2) = 2g-2 - c(\varphi_1) + c(\varphi_2) < 0$, so that necessarily $H^0(M, \mathcal{O}(\kappa\varphi_1^{-1}\varphi_2)) = 0$; this gives the desired result.

As an amusing sidelight, the cohomology class associated to an extension can be described in the following manner. Recall that $\mathcal{M}(\varphi_1\varphi_2^{-1}) = \mathcal{O}(\varphi_1\varphi_2^{-1}) \otimes \mathcal{M} \cong \mathcal{M}$, and hence that $H^1(M, \mathcal{M}(\varphi_1\varphi_2^{-1})) \cong H^1(M, \mathcal{M}) = 0$. Thus if $\sigma = (\sigma_{\alpha\beta}) \in Z^1(\mathfrak{U}, \mathcal{O}(\varphi_1\varphi_2^{-1}))$ is a cocycle describing a particular extension of φ_1 by φ_2, there will always exist a meromorphic cochain $h = (h_\alpha) \in C^0(\mathfrak{U}, \mathcal{M}(\varphi_1\varphi_2^{-1}))$ such that $\delta h = \sigma$. All extensions of meromorphic complex line bundles are thus trivial; and since all meromorphic line bundles are trivial, all meromorphic vector bundles are trivial - thus providing an extension to arbitrary Riemann surfaces of the remarks following Theorem 7, concerning the projective line. In terms of these meromorphic functions h_α, the meromorphic matrices

$$\Theta_\alpha(z_\alpha) = \begin{pmatrix} 1 & h_\alpha(z_\alpha) \\ 0 & 1 \end{pmatrix}$$

provide an explicit meromorphic change of coordinates, reducing the extension to the trivial extension; so $\Phi_{\alpha\beta} = \Theta_\alpha^{-1}\Theta_\beta$.

By the Serre duality theorem, $H^1(M, \mathcal{O}(\varphi_1\varphi_2^{-1}))$ is canonically dual to the space $\Gamma(M, \mathcal{O}(\kappa\varphi_1^{-1}\varphi_2)) = \Gamma(M, \mathcal{Q}^{1,0}(\varphi_1^{-1}\varphi_2))$; thus the cohomology class σ is canonically a linear functional on the vector space $\Gamma(M, \mathcal{Q}^{1,0}(\varphi_1^{-1}\varphi_2))$ of holomorphic differential forms which are sections of the bundle $\varphi_1^{-1}\varphi_2$. If $\omega = (\omega_\alpha) \in \Gamma(M, \mathcal{Q}^{1,0}(\varphi_1^{-1}\varphi_2))$, the value $\sigma(\omega)$ of the functional associated to σ on the section ω is given as follows, recalling §5(b) of last year's lectures. Letting $g = (g_\alpha) \in C^0(\mathcal{U}, \mathcal{E}^{0,0}(\varphi_1\varphi_2^{-1}))$ be any cochain such that $\delta g = \sigma$, the elements $\bar\partial g_\alpha$ define a global section $(\bar\partial g_\alpha) \in \Gamma(M, \mathcal{E}^{0,1}(\varphi_1\varphi_2^{-1}))$, and therefore $\bar\partial g_\alpha \wedge \omega_\alpha \in \Gamma(M, \mathcal{E}^{1,1})$; now

$$\sigma(\omega) = \iint_M \bar\partial g_\alpha \wedge \omega_\alpha .$$

Suppose that the covering $\mathcal{U} = \{U_\alpha\}$ is so chosen that the singularities of the meromorphic functions h_α are contained in disjoint open sets U_{α_i}, and are not contained in any intersection $U_\alpha \cap U_\beta$. For each i select a C^∞ function r_i in U_{α_i} which is identically 1 in each intersection $U_{\alpha_i} \cap U_\beta$, and which vanishes identically in a neighborhood of the singularities of h_{α_i} in U_{α_i}; and put $r_\alpha \equiv 1$ for the remaining sets of the covering $\mathfrak{U}$. Then in the construction of the Serre duality mapping it is possible to take $g_\alpha = r_\alpha h_\alpha$, so that $\sigma(\omega) = \iint_M \bar\partial(r_\alpha h_\alpha) \wedge \omega_\alpha$. However, except for the sets U_{α_i}, the function h_α is holomorphic and $r_\alpha \equiv 1$, so that $\bar\partial(r_\alpha h_\alpha) = 0$. Therefore, since $\bar\partial\omega_\alpha = 0$,

$$\sigma(\omega) = \sum_i \iint_{U_{\alpha_i}} \bar\partial(r_{\alpha_i} h_{\alpha_i}) \wedge \omega_{\alpha_i} = \sum_i \iint_{U_{\alpha_i}} \bar\partial(r_{\alpha_i} h_{\alpha_i} \omega_{\alpha_i})$$

$$= \sum_i \iint_{U_{\alpha_i}} d\,(r_{\alpha_i} h_{\alpha_i} \omega_{\alpha_i}) = \sum_i \int_{\partial U_{\alpha_i}} r_{\alpha_i} h_{\alpha_i} \omega_{\alpha_i},$$

by Stokes' theorem; and since $r_{\alpha_i} \equiv 1$ on ∂U_{α_i}, it follows that

$$\sigma(\omega) = \sum_i \int_{\partial U_{\alpha_i}} h_{\alpha_i} \omega_{\alpha_i}.$$

Note however that $h_\alpha \omega_\alpha \in C^0(\mathfrak{U}, \mathcal{M}^{1,0})$ and $\delta(\omega_\alpha h_\alpha) = \omega\sigma \in Z^1(\mathfrak{U}, \mathcal{Q}^{1,0})$, so that the singularities of $h_\alpha \omega_\alpha$ are globally defined; hence the total residue $\mathcal{R}(h_\alpha \omega_\alpha)$ of these singularities is well defined. Therefore,

$$\sigma(\omega) = 2\pi i\, \mathcal{R}(h_\alpha \omega_\alpha), \tag{8}$$

which is the result desired.

(b) Having classified the extensions of one line bundle by another, the next problem is that of determining which line bundles are possible as subbundles of a given vector bundle. This can be approached through an extension of the notion of the divisor of a meromorphic function, in the following manner.

If $F(z)$ is an m-tuple of meromorphic functions in a neighborhood of a point $p \in \mathbf{C}$, the order of $F(z)$ at the point p is that integer ν such that $(z-p)^{\nu}F(z)$ is an m-tuple of holomorphic functions with no common zeros in an open neighborhood of p; equivalently of course, ν is the least of the orders of the m component functions of $F(z)$. The order will be denoted by $\nu_p(F)$. Note that if Θ is a holomorphic invertible $m \times m$ matrix in an open neighborhood of p, then $\nu_p(F) = \nu_p(\Theta F)$. Now if $\Phi \in H^1(M, \mathcal{GL}(m, \mathcal{O}))$ is a vector bundle over M, and if $F \in \Gamma(M, \mathcal{M}(\Phi))$ is a meromorphic section of Φ, it is apparent that the order $\nu_p(F)$ of F at a point $p \in M$ is a well-defined concept. The divisor of the section F is then defined by

$$\mathcal{J}(F) = \sum_{p \in M} \nu_p(F) \cdot p ,$$

and is a divisor on the surface M in the usual sense. This divisor is associated to a unique line bundle $\varphi \in H^1(M, \mathcal{O}^*)$, in the sense that φ admits a meromorphic section $f \in \Gamma(M, \mathcal{M}(\varphi))$ with $\mathcal{J}(f) = \mathcal{J}(F)$.

Lemma 13. If Φ is a complex analytic vector bundle (of arbitrary rank) over a Riemann surface M, then a complex analytic line bundle φ is a subbundle of Φ if and only if, for every divisor $\mathcal{J}$ associated to φ, there is a meromorphic section $F \in \Gamma(M, \mathcal{M}(\Phi))$ such that $\mathcal{J} = \mathcal{J}(F)$.

Proof. If $\varphi \subset \Phi$, and if $\mathcal{J}$ is any divisor associated to φ, there is a meromorphic section $f \in \Gamma(M, \mathcal{M}(\varphi))$ such that $\mathcal{J} = \mathcal{J}(f)$; but by Lemma 11, f determines a section $F \in \Gamma(M, \mathcal{M}(\Phi))$, and since $\mathfrak{d}(F) = \mathfrak{d}(f) = \mathfrak{d}$, the implication in the one direction is demonstrated. Conversely, if φ is any line bundle over M associated to a divisor $\mathfrak{d}$, and if $F \in \Gamma(M, \mathcal{M}(\varphi))$ is a meromorphic section such that $\mathfrak{d} = \mathfrak{d}(F)$, then there is by Lemma 11 a subbundle $\psi \subset \Phi$ with a section f such that f determines F, hence $\mathfrak{d}(f) = \mathfrak{d}(F) = \mathfrak{d}$; but then necessarily $\psi = \varphi$, and the result is thereby proved.

Now consider an arbitrary complex analytic vector bundle $\Psi \in H^1(M, \mathcal{GL}(m, \mathcal{O}))$ over a compact Riemann surface; and suppose that Ψ is fully reduced, in the sense that it is defined by a cocycle $(\Psi_{\alpha\beta}) \in Z^1(\mathcal{U}, \mathcal{GL}(m, \mathcal{O}))$ of the form given in the Corollary to Theorem 10. If $F \in \Gamma(M, \mathcal{O}(\Psi))$ is a non-trivial holomorphic section, write

$$F = (F_\alpha) = \begin{pmatrix} f_{1\alpha} \\ f_{2\alpha} \\ \cdots \\ f_{m\alpha} \end{pmatrix} ;$$

and suppose that $(f_{r\alpha})$ is the last component of F that is not identically zero, so that $f_{r\alpha} \not\equiv 0$ but $f_{i\alpha} \equiv 0$ whenever $r < i \leqq m$. It is apparent then that $(f_{r\alpha}) \in \Gamma(M, \mathcal{O}(\psi_r))$, so that at least one of the diagonal line bundles in the reduction of Ψ must have a holomorphic section. Now if φ is any line bundle over M such that $\varphi \subset \Psi$, it follows that $1 \subset \varphi^{-1} \otimes \Psi$, where 1 denotes the trivial line bundle. By Lemma 13, the vector bundle $\varphi^{-1} \otimes \Psi$ admits therefore a holomorphic, non-trivial section F ; and therefore, as noted above, at least one of the diagonal bundles $\varphi^{-1} \otimes \psi_r$ must admit a non-trivial holomorphic section, so that

$c(\varphi^{-1} \otimes \psi_r) = c(\psi_r) - c(\varphi) \geqq 0$. Hence, for any line bundle $\varphi \subset \Psi$, it follows that $c(\varphi) \leqq \max_j c(\psi_j)$, where (ψ_j) are the diagonal line bundles in any fixed complete reduction of the bundle Ψ; or in other words, the Chern classes $c(\varphi)$ of line bundles φ which are subbundles of a fixed complex vector bundle Ψ, are bounded from above. Having made this observation, the divisor order of a complex analytic vector bundle $\Psi \in H^1(M, \mathcal{GL}(m, \mathcal{O}))$ over a compact Riemann surface M is the integer, denoted by div Ψ, defined by

$$\text{div}\ \Psi = \max_{\varphi \subset \Psi} c(\varphi) , \tag{9}$$

where $\{\varphi\}$ are the line bundles which are subbundles of Ψ.

The elementary properties of the divisor order of a vector bundle are easily established. First, recall from Theorem 11 of last year's lectures that for any line bundle $\varphi \in H^1(M, \mathcal{O}^*)$ over a compact Riemann surface M and any non-trivial section $f \in \Gamma(M, \mathcal{M}(\varphi))$ it follows that $c(\varphi) = \Sigma_{p \in M} \nu_p(f)$. It then follows immediately from Lemma 13 and the definitions, that for any vector bundle $\Psi \in H^1(M, \mathcal{GL}(m, \mathcal{O}))$ over a compact Riemann surface,

$$\text{div}\ \Psi = \max_{\{F \in \Gamma(M, \mathcal{M}(\Psi)),\ F \neq 0\}} \ \sum_{p \in M} \nu_p(F) . \tag{10}$$

Next, for line bundles φ, ξ and a vector bundle $\Psi \in H^1(M, \mathcal{GL}(m, \mathcal{O}))$, note that $\varphi \subset \Psi$ if and only if $\xi \otimes \varphi \subset \xi \otimes \Psi$; and therefore

$$\text{div}(\xi \otimes \Psi) = c(\xi) + \text{div}\ \Psi . \tag{11}$$

Finally, suppose that $\Psi \in H^1(M, \mathcal{GL}(m, \mathcal{O}))$ is fully reduced, in the sense that it is defined by a cocycle of the form given in the Corollary to Theorem 10; and let $\psi_1, \psi_2, \ldots, \psi_m$ be the diagonal line bundles in that reduction. (Thus Ψ is the result of successive extensions by the line bundles $\psi_1, \psi_2, \ldots, \psi_m$, in that order.) As noted in the paragraph immediately above, $c(\varphi) \leqq \max_j c(\psi_j)$ for any line bundle $\varphi \subset \Psi$; and therefore,

$$\text{(12)} \qquad \operatorname{div} \Psi \leqq \max_j c(\psi_j) .$$

In particular, it is clear that

$$\text{(13)} \qquad \operatorname{div}(\psi_1 \oplus \ldots \oplus \psi_m) = \max_j c(\psi_j) ,$$

since each bundle ψ_j is in this case a subbundle.

The following useful observation is also quite easy. Let $\Phi \in H^1(M, \mathcal{GL}(m, \mathcal{O}))$ be a complex vector bundle and $\varphi \in H^1(M, \mathcal{O}^*)$ be a complex line bundle on the compact Riemann surface M; then

$$\text{(14)} \qquad \Gamma(M, \mathcal{O}(\varphi^{-1} \otimes \Phi)) \neq 0 \ \text{ implies that } \ \operatorname{div} \Phi \geqq c(\varphi) .$$

For if $\varphi^{-1} \otimes \Phi$ has a non-trivial holomorphic section, it follows from Lemma 13 that there is a complex line bundle $\xi \subset \varphi^{-1} \otimes \Phi$ which has a non-trivial holomorphic section, hence such that $c(\xi) \geqq 0$; but then $\xi\varphi \subset \Phi$, and thus $\operatorname{div} \Phi \geqq c(\xi) + c(\varphi) \geqq c(\varphi)$.

<u>Lemma 14.</u> Let M be a compact Riemann surface of genus g; and let $\Phi \in H^1(M, \mathcal{GL}(m, \mathcal{O}))$ be a complex analytic vector bundle of rank m over M. The divisor order of Φ then satisfies the inequality

$$\operatorname{div} \Phi \geqq \frac{1}{m} c(\det \Phi) - g ;$$

and if Φ is decomposable, it further follows that

$$\operatorname{div} \Phi \geqq \frac{1}{m} c(\det \Phi) .$$

Proof. For any complex line bundle $\varphi \in H^1(M, \mathcal{O}^*)$ over the surface M, it follows from the Riemann-Roch theorem for complex vector bundles that

$$\begin{aligned} \dim H^0(M, \mathcal{O}(\varphi^{-1} \otimes \Phi)) &= \dim H^1(M, \mathcal{O}(\varphi^{-1} \otimes \Phi)) + \\ &\quad + c(\det(\varphi^{-1} \otimes \Phi)) + m(1-g) \\ &\geqq c(\det(\varphi^{-1} \otimes \Phi)) + m(1-g) \\ &\geqq c(\det \Phi) + m(1-g-c(\varphi)) . \end{aligned}$$

Whenever $c(\varphi) < \frac{1}{m} c(\det \Phi) + 1 - g$ it thus follows that $\dim H^0(M, \mathcal{O}(\varphi^{-1} \otimes \Phi)) > 0$, and therefore by (14) that $\operatorname{div} \Phi \geqq c(\varphi)$; since $c(\varphi)$ takes on arbitrary integral values, it follows that $\operatorname{div} \Phi \geqq n$ where n is any integer satisfying $n < \frac{1}{m} c(\det \Phi) + 1 - g$, and the first inequality follows easily from this. If Φ is decomposable as a direct sum $\Phi = \varphi_1 \oplus \ldots \oplus \varphi_m$, then since $c(\det \Phi) = c(\varphi_1) + \ldots + c(\varphi_m)$, necessarily $c(\varphi_j) \geqq \frac{1}{m} c(\det \Phi)$ for some index j; but from (13) it follows that $\operatorname{div} \Phi = \max_j c(\varphi_j) \geqq \frac{1}{m} c(\det \Phi)$, concluding the proof.

(c) Let us now return to the consideration of the special case of complex analytic vector bundles of rank 2 over a compact Riemann surface M. For any complex line bundle ξ, set

$$(15) \qquad \mathcal{V}(M,\xi) = \{\Phi \in H^1(M, \mathcal{GL}(2, \mathcal{O}))\,|\det \Phi = \xi\} .$$

Note that for any other line bundle η it follows that $\mathcal{V}(M,\xi) \otimes \eta = \mathcal{V}(M,\xi\eta^2)$; thus to describe all vector bundles, it suffices merely to describe the sets $\mathcal{V}(M,\xi_\nu)$ for fixed line bundles ξ_ν of Chern classes $c(\xi_\nu) = \nu$ taking just the values $\nu = -1, 0$. In particular, it suffices to describe the sets $\mathcal{V}(M,1)$ and $\mathcal{V}(M,\zeta_p^{-1})$, where ζ_p is the point bundle associated to a fixed base point $p \in M$. If the description behaves reasonably upon tensoring with line bundles, the classification problem will have been settled. In carrying out this classification, the set of vector bundles naturally decomposes into two components.

A complex vector bundle $\Phi \in H^1(M, \mathfrak{GL}(2, \mathcal{O}))$ is called <u>stable</u> if

$$c(\varphi) < \frac{1}{2}\, c(\det \Phi)$$

for every line bundle $\varphi \subset \Phi$; otherwise, Φ is called <u>unstable</u>. (This terminology and classification was introduced by D. Mumford; see for instance Proc. Int. Congress of Math., Stockholm (1962), 526-530.) Note for any line bundle ψ that $\varphi \subset \psi \otimes \Phi$ if and only if $\psi^{-1}\varphi \subset \Phi$. Now if Φ is stable and $\varphi \subset \psi \otimes \Phi$, it follows that $c(\psi^{-1}\varphi) < \frac{1}{2}\, c(\det \Phi)$; hence $c(\varphi) = c(\psi^{-1}\varphi) + c(\psi) <$ $< \frac{1}{2}\, c(\det \Phi) + c(\psi) = \frac{1}{2}\, c(\det(\psi \otimes \Phi))$, so that $\psi \otimes \Phi$ is also stable. The notion of stability is therefore well-behaved under tensoring with line bundles. Let $\mathcal{V}'(M,\xi)$ denote the set of stable vector bundles in $\mathcal{V}(M,\xi)$, and $\mathcal{V}''(M,\xi)$ denote the set of unstable vector bundles in $\mathcal{V}(M,\xi)$; so $\mathcal{V}(M,\xi)$ can be decomposed into the disjoint union

$$\mathcal{V}(M,\xi) = \mathcal{V}'(M,\xi) \cup \mathcal{V}''(M,\xi) .$$

Then $\mathcal{V}'(M,\xi) \otimes \eta = \mathcal{V}'(M,\xi\eta^2)$ and $\mathcal{V}''(M,\xi) \otimes \eta = \mathcal{V}''(M,\xi\eta^2)$ for any line bundle η. Note that $\Phi \in \mathcal{V}(M,\xi)$ is stable precisely when $\operatorname{div} \Phi < \frac{1}{2} c(\xi)$; for by definition, $\operatorname{div} \Phi = \max_{\varphi \subset \Phi} c(\Phi)$. Then

$$(16) \qquad \begin{cases} \mathcal{V}'(M,\xi) = \{\Phi \in \mathcal{V}(M,\xi) \mid \operatorname{div} \Phi < \frac{1}{2} c(\xi)\} \\ \mathcal{V}''(M,\xi) = \{\Phi \in \mathcal{V}(M,\xi) \mid \operatorname{div} \Phi \geqq \frac{1}{2} c(\xi)\} \end{cases}$$

<u>Lemma 15.</u> Let M be a compact Riemann surface, and consider a vector bundle $\Phi \in \mathcal{V}(M,\xi)$.

(i) If $\operatorname{div} \Phi > \frac{1}{2} c(\xi)$, there is a unique line bundle φ_1 such that $\varphi_1 \subset \Phi$ and $c(\varphi_1) = \operatorname{div} \Phi$.

(ii) If $\operatorname{div} \Phi = \frac{1}{2} c(\xi)$, and Φ is indecomposable, there is a unique line bundle φ_1 such that $\varphi_1 \subset \Phi$ and $c(\varphi_1) = \operatorname{div} \Phi$.

Proof. (i) Suppose that $\operatorname{div} \Phi > \frac{1}{2} c(\xi)$; let $\varphi_1 \subset \Phi$ be a subbundle for which $c(\varphi_1) = \operatorname{div} \Phi$, and let $\varphi_2 = \Phi/\varphi_1$ be the quotient bundle. Thus the vector bundle Φ has a full reduction, in the sense of the Corollary to Theorem 10, where φ_1, φ_2' are the diagonal line bundles. If $\varphi \subset \Phi$ is another subbundle for which $c(\varphi) = \operatorname{div} \Phi$, the vector bundle $\varphi^{-1} \otimes \Phi$ admits a non-trivial holomorphic section, by Lemma 13; but then, as in the discussion on page 79, at least one of the diagonal bundles $\varphi^{-1}\varphi_1$, $\varphi^{-1}\varphi_2$ must also admit a non-trivial holomorphic section. Since $c(\varphi^{-1}\varphi_2) = c(\varphi_1^{-1}\varphi_2) = -c(\varphi_1) + c(\varphi_2) = -2c(\varphi_1) + c(\det \Phi) =$ $= -2c(\varphi_1) + c(\xi) = -2 \operatorname{div} \Phi + c(\xi) < 0$, that bundle has no non-trivial holomorphic sections; so the bundle $\varphi^{-1}\varphi_1$ must have such sections. Since $c(\varphi^{-1}\varphi_1) = 0$, necessarily $\varphi^{-1}\varphi_1 = 1$, and so $\varphi = \varphi_1$, as desired.

(ii) If $\operatorname{div}\Phi = \frac{1}{2}c(\xi)$, the argument begins as in part (i); the only change is that now $c(\varphi^{-1}\varphi_2) = 0$. To continue from that point, let $F = \begin{pmatrix} f_1 \\ f_2 \end{pmatrix}$ be a holomorphic non-trivial section of $\varphi^{-1}\otimes\Phi$; it is merely necessary to show that $f_2 \equiv 0$, and conclude the argument as in part (i). Suppose, contrariwise, that f_2 is not identically zero; since f_2 is a section of $\varphi^{-1}\varphi_2$, and $c(\varphi^{-1}\varphi_2) = 0$, necessarily $\varphi^{-1}\varphi_2 = 1$. In terms of a suitable coordinate covering $\mathfrak{U} = \{U_\alpha\}$, the bundle $\Psi = \varphi^{-1}\otimes\Phi$ can be defined by a cocycle $(\Psi_{\alpha\beta})$ of the form

$$\Psi_{\alpha\beta} = \begin{pmatrix} \psi_{\alpha\beta} & \lambda_{\alpha\beta} \\ 0 & 1 \end{pmatrix} ;$$

and it has then a holomorphic section of the form

$$F = (F_\alpha) = \begin{pmatrix} f_{1\alpha} \\ c \end{pmatrix},$$

where c is a non-zero constant. The functions $(f_{1\alpha})$ are thus holomorphic in U_α , and satisfy

$$f_{1\alpha}(p) = \psi_{\alpha\beta}(p)f_{1\beta}(p) + c\lambda_{\alpha\beta}(p) \quad \text{for} \quad p \in U_\alpha \cap U_\beta .$$

But then the cocycle $(c\lambda_{\alpha\beta})$, hence the cocycle $(\lambda_{\alpha\beta})$ itself, must be cohomologous to zero in $H^1(M, \mathcal{O}(\psi))$; as in Theorem 13, the bundle $\Psi = \varphi^{-1}\otimes\Phi$, is then decomposable, so that Φ is itself decomposable. This contradiction serves to conclude the proof.

Now let $\mathcal{V}^0(M,\xi)$ denote the subset of $\mathcal{V}(M,\xi)$ consisting of decomposable bundles. It follows immediately from Lemma 14 that $\operatorname{div}\Phi \geqq \frac{1}{2}c(\xi)$ whenever $\Phi \in \mathcal{V}^0(M,\xi)$, hence that every decomposable bundle is unstable; that is to say

$$\mathcal{V}^{0}(M,\xi) \subset \mathcal{V}''(M,\xi) .$$

The set $\mathcal{V}^0(M,\xi)$ can be described quite easily. Note firstly that $\varphi_1 \oplus \varphi_2 = \varphi_1' \oplus \varphi_2' = \Phi \in \mathcal{V}^0(M,\xi)$ if and only if either $\varphi_1 = \varphi_1'$ or $\varphi_1 = \varphi_2'$. (For suppose that φ_1 has the largest Chern class of these four bundles. Since $\varphi_1 \subset \Phi$, it follows that $1 \subset \varphi_1^{-1} \otimes \Phi$; hence by Lemma 13, the bundle $\varphi_1^{-1} \otimes \Phi$ admits a non-trivial holomorphic section. Then one of the two bundles $\varphi_1^{-1}\varphi_1'$ or $\varphi_1^{-1}\varphi_2'$ admits a holomorphic non-trivial section; but since $c(\varphi_1^{-1}\varphi_j') \leqq 0$, it is clear that $\varphi_1 = \varphi_j'$ for either $j = 1$ or $j = 2$.) Now consider the mapping

$$H^1(M, \mathcal{O}^*) \longrightarrow \mathcal{V}^0(M,\xi)$$

defined by

$$\varphi_1 \in H^1(M, \mathcal{O}^*) \longrightarrow (\varphi_1 \oplus \varphi_1^{-1}\xi) \in \mathcal{V}^0(M,\xi) .$$

It is evident that two distinct line bundles φ_1, φ_2 have the same image under this mapping precisely when $\varphi_2 = \varphi_1^{-1}\xi$; thus $\mathcal{V}^0(M,\xi)$ is naturally identified with the quotient space of $H^1(M, \mathcal{O}^*)$ under the involution $\varphi \longrightarrow \varphi^{-1}\xi$. The group $H^1(M, \mathcal{O}^*)$ has the structure of a complex analytic manifold of dimension g (with infinitely many components), as noted in last year's lectures; and the involution $\varphi \longrightarrow \varphi^{-1}\xi$ is an analytic mapping. Therefore $\mathcal{V}^0(M,\xi)$ also has the structure of a complex analytic variety of dimension g. (This variety has singularities at the fixed points of the involution.)

For an indecomposable unstable vector bundle in $\mathcal{V}(M,\xi)$, that is to say, for a bundle $\Phi \in \mathcal{V}''(M,\xi) - \mathcal{V}^0(M,\xi)$, it follows directly from Lemma 15 that there is a unique line bundle φ such that $\varphi \subset \Phi$ and $c(\varphi) = \text{div}\,\Phi$. Therefore, letting

(17) $\mathcal{V}''(M,\xi;\varphi) = \{\Phi \in \mathcal{V}''(M,\xi) - \mathcal{V}^0(M,\xi) \mid \varphi \subset \Phi \text{ and } c(\varphi) = \text{div}\,\Phi\}$,

the sets (17) are disjoint, for different bundles ξ, φ ; so <u>the set of unstable vector bundles over</u> M <u>can be written as the disjoint union</u>

(18) $$\mathcal{V}''(M,\xi) = \mathcal{V}^0(M,\xi) \cup \Big(\bigcup_{\substack{\varphi \\ c(\varphi) \geqq \frac{1}{2}c(\xi)}} \mathcal{V}''(M,\xi;\varphi) \Big)$$

for every compact Riemann surface M ; the union is over all the line bundles φ satisfying $c(\varphi) \geqq \frac{1}{2}c(\xi)$. There remains the problem of describing in more detail the sets $\mathcal{V}''(M,\xi;\varphi)$.

<u>Theorem 14.</u> On a compact Riemann surface M of genus g there is a natural one-to-one correspondence between the elements of $\mathcal{V}''(M,\xi;\varphi)$, for any complex line bundles ξ,φ such that $c(\varphi) \geqq \frac{1}{2}c(\xi)$, and the points of the complex projective space $\mathbb{P}^n$ of dimension $n = \dim H^1(M, \mathcal{O}(\varphi^2\xi^{-1})) - 1$.

Proof. Given the line bundles ξ,φ, subject to the condition that $c(\varphi) \geqq \frac{1}{2}c(\xi)$, let $\Phi \in H^1(M, \mathcal{B}\mathcal{L}(2, \mathcal{O}))$ be any extension of φ by $\varphi_2 = \xi\varphi^{-1}$. Since $c(\varphi_2) = c(\xi) - c(\varphi) \leqq \frac{1}{2}c(\xi)$, it follows readily from (12) that $\text{div}\,\Phi = c(\varphi) \geqq \frac{1}{2}c(\xi) = \frac{1}{2}c(\det\Phi)$, and therefore Φ is unstable. That is, then,

$$\mathcal{V}''(M,\xi;\varphi) = \{\Phi \in H^1(M, \mathcal{B}\mathcal{L}(2, \mathcal{O})) \mid \varphi \subset \Phi \text{ and } \det\Phi = \xi\} \cap \{\text{indecomposable vector bundles}\}.$$

The set of equivalence classes of extensions of φ by $\varphi_2 = \xi\varphi^{-1}$

was determined in Theorem 13; but this classification cannot be applied directly, since for the present purposes the weaker equivalence of vector bundles is of interest, rather than the equivalence of extensions. (That is to say, there is no restriction the way φ is realized as a subbundle of Φ .)

However, suppose that Φ is any indecomposable extension of φ by $\varphi_2 = \xi\varphi^{-1}$; and choose a cocycle $\Phi_{\alpha\beta} \in Z^1(\mathfrak{U}, \mathfrak{L}(2, \mathcal{O}))$ representing the vector bundle Φ , in terms of some covering $\mathfrak{U} = \{U_\alpha\}$ of M . This cocycle can be taken in the form

$$(19) \qquad \Phi_{\alpha\beta} = \begin{pmatrix} \varphi_{\alpha\beta} & \lambda_{\alpha\beta} \\ 0 & \varphi_{2\alpha\beta} \end{pmatrix},$$

where $(\varphi_{\alpha\beta}) \in Z^1(\mathfrak{U}, \mathcal{O}^*)$ represents φ and $(\varphi_{2\alpha\beta}) \in Z^1(\mathfrak{U}, \mathcal{O}^*)$ represents $\varphi_2 = \xi\varphi^{-1}$; and, as in the proof of Theorem 13, the elements $\lambda_{\alpha\beta}$ are arbitrary analytic functions in $U_\alpha \cap U_\beta$, subject only to the condition that the elements $\sigma_{\alpha\beta}(z_\beta) = \varphi_{\alpha\beta}^{-1}(p)\lambda_{\alpha\beta}(p)$, for $z_\beta = z_\beta(p)$, form a cocycle $(\sigma_{\alpha\beta}) \in Z^1(\mathfrak{U}, \mathcal{O}(\varphi\varphi_2^{-1})) = Z^1(\mathfrak{U}, \mathcal{O}(\varphi^2\xi^{-1}))$. If Φ' is another extension of φ by φ_2 , after passing to a refinement of the covering if necessary, Φ' will be defined by a cocycle $(\Phi'_{\alpha\beta})$ of the form (19), but with some other functions $\lambda'_{\alpha\beta}$; and let $(\sigma'_{\alpha\beta}) \in Z^1(\mathfrak{U}, \mathcal{O}(\varphi^2\xi^{-1}))$ be the cocycle associated to $(\lambda'_{\alpha\beta})$. The bundles Φ and Φ' are the same (that is, are analytically equivalent) if and only if, after passing to another refinement of the covering perhaps, there are functions $\Theta_\alpha \in GL(2, \mathcal{O}_{U_\alpha})$ such that

$$(20) \qquad \Phi'_{\alpha\beta} = \Theta_\alpha\Phi_{\alpha\beta}\Theta_\beta^{-1} \quad \text{in} \quad U_\alpha \cap U_\beta .$$

Writing the matrices Θ_α explicitly as

$$\Theta_\alpha = \begin{pmatrix} f_{11\alpha} & f_{12\alpha} \\ f_{21\alpha} & f_{22\alpha} \end{pmatrix},$$

equation (20) takes the form

$$(21)\quad \begin{pmatrix} \varphi_{\alpha\beta} & \lambda'_{\alpha\beta} \\ 0 & \varphi_{2\alpha\beta} \end{pmatrix} \begin{pmatrix} f_{11\beta} & f_{12\beta} \\ f_{21\beta} & f_{22\beta} \end{pmatrix} = \begin{pmatrix} f_{11\alpha} & f_{12\alpha} \\ f_{21\alpha} & f_{22\alpha} \end{pmatrix} \begin{pmatrix} \varphi_{\alpha\beta} & \lambda_{\alpha\beta} \\ 0 & \varphi_{2\alpha\beta} \end{pmatrix}.$$

Considering at first the functions $(f_{21\alpha})$, it follows from (21) that $\varphi_{2\alpha\beta}f_{21\beta} = f_{21\alpha}\varphi_{\alpha\beta}$, hence that $(f_{21\alpha}) \in \Gamma(M, \mathcal{O}(\varphi\varphi_2^{-1})) = \Gamma(M, \mathcal{O}(\varphi^2\xi^{-1}))$. Note however that $c(\varphi^2\xi^{-1}) = 2c(\varphi) - c(\xi) \leqq 0$; so the section $(f_{21\alpha})$ can be non-trivial only when $c(\varphi) = \frac{1}{2}c(\xi)$, so $c(\varphi\varphi_2^{-1}) = c(\varphi^2\xi^{-1}) = 0$. But if $c(\varphi\varphi_2^{-1}) = 0$ and $\varphi\varphi_2^{-1}$ has a non-trivial holomorphic section, necessarily $\varphi\varphi_2^{-1} = 1$; and letting φ and φ_2 both be defined by the same cocycle $(\varphi_{\alpha\beta})$, the section $f_{21\alpha}$ becomes a globally defined holomorphic function on the compact Riemann surface M, hence a non-zero constant c. It then further follows from (21) that $\varphi_{2\alpha\beta}f_{22\beta} = c\lambda_{\alpha\beta} + f_{22\alpha}\varphi_{2\alpha\beta}$, or equivalently that

$$f_{22\beta} - f_{22\alpha} = c\varphi_{2\alpha\beta}^{-1}\lambda_{\alpha\beta}.$$

This last equation shows that $c\cdot(\varphi_{2\alpha\beta}^{-1}\lambda_{\alpha\beta}) = 0$ in $H^1(M, \mathcal{O}) = H^1(M, \mathcal{O}(\varphi\varphi_2^{-1}))$, and since $c \neq 0$, necessarily $(\varphi_{2\alpha\beta}^{-1}\lambda_{\alpha\beta}) = 0$ as well; but then, recalling the correspondence of Theorem 13, it follows that the bundle Φ is decomposable, a contradiction. Therefore it is necessary that $f_{21\alpha} \equiv 0$. Using this fact, equation (21) clearly reduces to the following equations:

$$(22) \qquad \begin{cases} f_{11\alpha} = f_{11\beta} \; ; \\ f_{22\alpha} = f_{22\beta} \; ; \\ \varphi_{\alpha\beta} f_{12\beta} + \lambda'_{\alpha\beta} f_{22\beta} = f_{11\alpha}\lambda_{\alpha\beta} + f_{12\alpha}\varphi_{2\alpha\beta} \; . \end{cases}$$

The first equation in (22) shows that the various functions $(f_{11\alpha})$ define a global holomorphic function a_1 over M, which must be constant since M is compact; and similarly, the functions $f_{22\alpha}$ reduce to a constant a_2. Necessarily $a_1 a_2 \neq 0$, of course. Thus (21) reduces to the condition that

$$(23) \qquad a_1\lambda_{\alpha\beta} - a_2\lambda'_{\alpha\beta} = \varphi_{\alpha\beta} f_\beta - f_\alpha \varphi_{2\alpha\beta} \; ,$$

where a_1, a_2 are non-zero constants, and $f_\alpha = f_{12\alpha}$ are holomorphic functions in U_α. Again, as in Theorem 13, consider the sections $(f_\alpha(z_\alpha)) \in \Gamma(U_\alpha, \mathcal{O}(\varphi\varphi_2^{-1}))$ associated to the functions $f_\alpha(p) \in \Gamma(U_\alpha, \mathcal{O})$, defined by $f_\alpha(z_\alpha(p)) = f_\alpha(p)$; and multiplying (23) through by $\varphi_{\alpha\beta}^{-1}$, it finally reduces to the equation

$$(24) \qquad a_1\sigma_{\alpha\beta}(z_\beta) - a_2\sigma'_{\alpha\beta}(z_\beta) = f_\beta(z_\beta) - f_\alpha(z_\beta) \quad \text{in } U_\alpha \cap U_\beta \; .$$

The vector bundles Φ and Φ' defined by the cocycles $(\sigma_{\alpha\beta})$ and $(\sigma'_{\alpha\beta})$ in $Z^1(\mathfrak{U}, \mathcal{O}(\varphi\varphi_2^{-1})) = Z^1(\mathfrak{U}, \mathcal{O}(\varphi^2\xi^{-1}))$ are thus equivalent if and only if there are non-zero constants a_1, a_2 and a zero-cochain $(f_\alpha) \in C^0(\mathfrak{U}, \mathcal{O}(\varphi^2\xi^{-1}))$ such that (24) holds, that is, if and only if there are non-zero constants a_1, a_2, such that $a_1\sigma - a_2\sigma' = 0$ in $H^1(M, \mathcal{O}(\varphi^2\xi^{-1}))$. Since the bundles Φ and Φ' are indecomposable, the cohomology classes σ and σ' are non-zero, as in Theorem 13; and thus σ and σ' can be considered as representing elements in the projective space

$\mathbb{P}^n = H^1(M, \mathcal{O}(\varphi^2\xi^{-1}))/\mathbf{C}^*$. The final result is that bundles associated to σ , σ' are the same if and only if σ and σ' represent the same point in the projective space $\mathbb{P}^n$, where $n = \dim H^1(M, \mathcal{O}(\varphi^2\xi^{-1})) - 1$; that concludes the proof of the theorem.

In one sense, formula (18) and Theorem 14 provide a descriptive classification of the sets $\mathcal{V}''(M,\xi)$ of unstable vector bundles over M . All the components in (14) are complex analytic space of dimension $\leqq g$. (By the Serre duality theorem, $H^1(M, \mathcal{O}(\varphi^2\xi^{-1})) \cong H^0(M, \mathcal{O}(\kappa\varphi^{-2}\xi))$; and since $c(\kappa\varphi^{-2}\xi) = c(\kappa) - 2c(\varphi) + c(\xi) \leqq 2g - 2$, it follows from the Riemann-Roch theorem (recall formula (14) on page 113 of last year's lectures) that $\dim H^1(M, \mathcal{O}(\varphi^2\xi^{-1})) \leqq g$. Therefore $\dim \mathcal{V}''(M,\xi;\varphi) \leqq g - 1$.) The question arises, whether the entire set $\mathcal{V}''(M,\xi)$ can be given the structure of a complex analytic variety, with all these components imbedded as complex analytic subvarieties; and any complete classification theory must provide an answer to this question. We shall return to this question later.

(d) This approach to the descriptive classification of complex vector bundles does not seem to be of use for stable bundles. For in the stable case, there is no uniqueness result analogous to Lemma 15; indeed, in the extreme case, exactly the opposite of Lemma 15 is true. Recall from the definitions and Lemma 14 that if $\Phi \in \mathcal{V}(M,\xi)$ is a stable bundle, then $\frac{1}{2}c(\xi) - g \leqq \operatorname{div}\Phi < \frac{1}{2}c(\xi)$; the lowest possible value of $\operatorname{div}\Phi$ is thus $[\frac{c(\xi)+1}{2}] - g$, where the square brackets denote the largest integer function.

Lemma 16. Let M be a compact Riemann surface of genus g, and let $\Phi \in \mathcal{V}(M,\xi)$ be a stable vector bundle such that

$$\operatorname{div} \Phi = \left[\frac{c(\xi)+1}{2}\right] - g .$$

Then for any line bundle $\varphi \in H^1(M, \mathcal{O}^*)$ such that $c(\varphi) = \operatorname{div} \Phi$, it follows that $\varphi \subset \Phi$.

Proof. To show that $\varphi \subset \Phi$, where φ is a line bundle such that $c(\varphi) = \operatorname{div} \Phi$, it suffices to show that the vector bundle $\varphi^{-1} \otimes \Phi$ has a non-trivial holomorphic section. For if $\varphi^{-1} \otimes \Phi$ has such a section F, there is by Lemma 11 a line bundle $\eta \subset \varphi^{-1} \otimes \Phi$ with a non-trivial holomorphic section $f \in \Gamma(M, \mathcal{O}(\eta))$ inducing the section F; and of course, $c(\eta) \geqq 0$. Now since $\varphi\eta \subset \Phi$, it follows that $c(\varphi) = \operatorname{div} \Phi \geqq c(\eta\varphi) = c(\eta) + c(\varphi)$, so that actually $c(\eta) = 0$; but since η has a non-trivial holomorphic section, necessarily $\eta = 1$, and $\varphi \subset \Phi$.

Given the vector bundle $\Phi \in \mathcal{V}(M,\xi)$ satisfying the hypotheses of the Lemma, let φ_1 be a line bundle such that $\varphi_1 \subset \Phi$ and $c(\varphi_1) = \operatorname{div} \Phi$; and let $\varphi_2 = \Phi/\varphi_1 \cong \xi\varphi_1^{-1}$ be the quotient bundle. Let φ be any line bundle with $c(\varphi) = \operatorname{div} \Phi = c(\varphi_1)$ and $\varphi \neq \varphi_1$; thus it is required to show that $\varphi^{-1} \otimes \Phi$ has a non-trivial holomorphic section. In terms of a suitable coordinate covering $\mathfrak{U} = \{U_\alpha\}$ of the surface M, the bundle Φ can be defined by a cocycle $(\Phi_{\alpha\beta}) \in Z^1(\mathfrak{U}, \mathcal{GL}(2, \mathcal{O}))$ of the form

$$\Phi_{\alpha\beta} = \begin{pmatrix} \varphi_{1\alpha\beta} & \lambda_{\alpha\beta} \\ 0 & \varphi_{2\alpha\beta} \end{pmatrix},$$

where $(\varphi_{j\alpha\beta}) \in Z^1(\mathcal{U}, \mathcal{O}^*)$ defines the line bundle φ_j ; and the line bundle φ can be defined by a cocycle $(\varphi_{\alpha\beta}) \in Z^1(\mathcal{U}, \mathcal{O}^*)$. A section $F \in \Gamma(M, \mathcal{O}(\varphi^{-1} \otimes \Phi))$ then consists of pairs of functions $f_{1\alpha}, f_{2\alpha} \in \Gamma(U_\alpha, \mathcal{O})$ satisfying

$$(25) \quad \begin{cases} f_{1\alpha}(p) = \varphi_{\alpha\beta}^{-1}(p)\varphi_{1\alpha\beta}(p)f_{1\beta}(p) + \varphi_{\alpha\beta}^{-1}(p)\lambda_{\alpha\beta}'(p)f_{2\beta}(p) \\ f_{2\alpha}(p) = \varphi_{\alpha\beta}^{-1}(p)\varphi_{2\alpha\beta}(p)f_{2\beta}(p) \end{cases}$$

whenever $p \in U_\alpha \cap U_\beta$. The functions $(f_{2\alpha})$ form a section $f_2 \in \Gamma(M, \mathcal{O}(\varphi^{-1}\varphi_2))$. For each such section f_2 , the functions $\varphi_{\alpha\beta}^{-1}(p)\lambda_{\alpha\beta}(p)f_{2\beta}(p)$ are readily seen to form a cocycle in $Z^1(\mathcal{U}, \mathcal{O}(\varphi^{-1}\varphi_1))$; and the condition that there exist holomorphic functions $(f_{1\alpha})$ satisfying the first line in (25), is just that this cocycle be cohomologous to zero. (Recall in this connection the proof of Theorem 13.) The mapping which associates to each section f_2 this cocycle thus yields a linear mapping

$$(26) \qquad \Gamma(M, \mathcal{O}(\varphi^{-1}\varphi_2)) \longrightarrow H^1(M, \mathcal{O}(\varphi^{-1}\varphi_1)) ;$$

and a given section $f_2 \in \Gamma(M, \mathcal{O}(\varphi^{-1}\varphi_2))$ can be extended to a section $F = \binom{f_1}{f_2} \in \Gamma(M, \mathcal{O}(\varphi^{-1} \otimes \Phi))$ precisely when f_2 is in the kernel of the linear mapping (26). In particular, to show that there is a non-trivial holomorphic section $F \in \Gamma(M, \mathcal{O}(\varphi^{-1} \otimes \Phi))$, it clearly suffices to show that there is a non-trivial element in the kernel of (26); and that will obviously be the case whenever $\dim \Gamma(M, \mathcal{O}(\varphi^{-1}\varphi_2)) > \dim H^1(M, \mathcal{O}(\varphi^{-1}\varphi_1))$. The proof of the Lemma is thereby reduced to demonstrating this last inequality.

By the Riemann-Roch theorem for complex line bundles (page 111 and following, in last year's lectures),

$$\dim \Gamma(M, \mathcal{O}(\varphi^{-1}\varphi_2)) \geqq c(\varphi^{-1}\varphi_2) - (g-1) = c(\xi) - 2 \operatorname{div} \Phi + 1 - g .$$

From the Serre duality theorem for line bundles, $H^1(M, \mathcal{O}(\varphi^{-1}\varphi_1)) \cong \Gamma(M, \mathcal{O}(\kappa\varphi\varphi_1^{-1}))$; and since $c(\kappa\varphi\varphi_1^{-1}) = c(\kappa) + c(\varphi) - c(\varphi_1) = 2g - 2$ but $\kappa\varphi\varphi_1^{-1} \neq \kappa$ (since by assumption $\varphi \neq \varphi_1$), it follows again from the Riemann-Roch theorem that

$$\dim H^1(M, \mathcal{O}(\varphi^{-1}\varphi_1)) = \dim \Gamma(M, \mathcal{O}(\kappa\varphi\varphi_1^{-1})) = g-1 .$$

Therefore

$$\dim \Gamma(M, \mathcal{O}(\varphi^{-1}\varphi_2)) - \dim H^1(M, \mathcal{O}(\varphi^{-1}\varphi_1)) \geqq c(\xi) - 2 \operatorname{div} \Phi + 2 - 2g > 0,$$

which concludes the proof.

We shall return to the classification of stable bundles later, using rather a different approach.

(e) It is perhaps of some interest to examine more closely the preceding approach to the classification of vector bundles, over Riemann surfaces of low genus. Consider firstly a compact Riemann surface $M = \mathbb{P}$ of genus zero. The stable bundles $\mathcal{V}'(M,\xi) \subset \mathcal{V}(M,\xi)$ are characterized by $\frac{1}{2}c(\xi) > \operatorname{div} \Phi \geqq \frac{1}{2}c(\xi)$, according to Lemma 14 and the definitions; so $\mathcal{V}'(M,\xi) = \emptyset$. Also, for any line bundle φ_1 the bundles $\mathcal{V}''(M,\xi;\varphi)$ are in one-to-one correspondence with the points of the projective space $\mathbb{P}^n$, where $n = \dim H^1(M, \mathcal{O}(\varphi^2\xi^{-1})) - 1$; but by the Serre duality theorem $H^1(M, \mathcal{O}(\varphi^2\xi^{-1})) \cong \Gamma(M, \mathcal{O}(\kappa\varphi^{-2}\xi))$, and since $c(\kappa\varphi^{-2}\xi) = -2 - 2c(\varphi) + c(\xi) \leqq -2$, it follows from the Riemann-Roch theorem that $\Gamma(M, \mathcal{O}(\kappa\varphi^{-2}\xi)) = 0$. Thus $n = -1$, and $\mathcal{V}''(M,\xi;\varphi) = \emptyset$. Altogether then,

$$\mathcal{V}(\mathbb{P},\xi) = \mathcal{V}^0(\mathbb{P},\xi) , \tag{27}$$

that is, <u>all vector bundles in $\mathcal{V}(\mathbb{P},\xi)$ are decomposable.</u> It is clear that this extends readily to vector bundles of all ranks over $\mathbb{P}$. (This observation was first made by A. Grothendieck in Amer. Jour. Math. 79(1957), 121-138.)

Next consider a compact Riemann surface of genus $g = 1$. The stable bundles $\mathcal{V}'(M,\xi) \subset \mathcal{V}(M,\xi)$ are then characterized by $\frac{1}{2}c(\xi) > \operatorname{div}\Phi \geqq \frac{1}{2}c(\xi) - 1$; thus necessarily $\operatorname{div}\Phi = [\frac{c(\xi)+1}{2}] - 1$, and all stable bundles belong to the extreme class discussed in Lemma 16. (That lemma can then be applied to describe this particular set of bundles in more detail; this was carried out by M. F. Atiyah in Proc. London Math. Soc. 7(1957), 414-452. We shall not pursue this approach further here, preferring a different technique for the treatment of stable bundles.) Again, for any line bundle φ, there is a natural one-to-one correspondence $\mathcal{V}''(M,\xi;\varphi) \leftrightarrow \mathbb{P}^n$ where $n = \dim H^1(M, \mathcal{O}(\varphi^2\xi^{-1})) - 1$; here $H^1(M,\mathcal{O}(\varphi^2\xi^{-1})) \cong \Gamma(M, \mathcal{O}(\kappa\varphi^{-2}\xi)) = \Gamma(M, \mathcal{O}(\varphi^{-2}\xi))$, since $\kappa = 1$, and from $c(\varphi^{-2}\xi) = c(\xi) - 2c(\varphi)$ and the Riemann-Roch theorem, it follows that $\Gamma(M, \mathcal{O}(\varphi^{-2}\xi)) = 0$ except when $\varphi^2 = \xi$. When $\varphi^2 = \xi$, of course $\Gamma(M, \mathcal{O}(\varphi^{-2}\xi)) = \Gamma(M, \mathcal{O}) = \mathbb{C}$. Therefore $\mathcal{V}''(M,\xi;\varphi) = \emptyset$ whenever $\varphi^2 \neq \xi$, while $\mathcal{V}''(M,\xi;\varphi)$ consists of a single element whenever $\varphi^2 = \xi$. Thus, <u>over a compact Riemann surface</u> M <u>of genus</u> $g = 1$,

(28) $$\mathcal{V}(M,\xi) = \mathcal{V}'(M,\xi) \cup \mathcal{V}^0(M,\xi) \cup \bigcup_{\varphi^2=\xi} \mathcal{V}''(M,\xi;\varphi),$$

where

$$\mathcal{V}'(M,\xi) = \{\Phi \in \mathcal{V}(M,\xi) \mid \operatorname{div}\Phi = [\tfrac{c(\xi)+1}{2}] - 1\}$$

and

$\mathcal{V}''(M,\xi;\varphi)$ has a single element whenever $\varphi^2 = \xi$.

§6. Flat vector bundles.

(a) Over an arbitrary Riemann surface M, the constant sheaf $GL(m,\mathbf{C}) \times M$ can be viewed as the subsheaf of the sheaf $\mathcal{GL}(m, \mathcal{O})$ of germs of complex analytic mappings from M into the complex Lie group $GL(m,\mathbf{C})$, consisting of germs of locally constant mappings. The inclusion mapping of sheaves

$$i: GL(m,\mathbf{C}) \longrightarrow \mathcal{GL}(m, \mathcal{O})$$

induces a mapping of the cohomology sets

$$(1) \qquad i^*: H^1(M, GL(m,\mathbf{C})) \longrightarrow H^1(M, \mathcal{GL}(m, \mathcal{O})) .$$

The elements of the set $H^1(M, GL(m,\mathbf{C}))$ will be called flat complex vector bundles of rank m over the Riemann surface M. A complex analytic vector bundle which lies in the image of i^* will be said to possess a flat representative; for such a complex analytic vector bundle can be defined by a cocycle $(\Phi_{\alpha\beta}) \in Z^1(\mathfrak{U}, \mathcal{GL}(m, \mathcal{O}))$ in which all of the functions $\Phi_{\alpha\beta}$ are constants. The set of all flat complex vector bundles mapping onto a given complex analytic vector bundle $\Phi \in H^1(M, \mathcal{GL}(m, \mathcal{O}))$ will be called the set of flat representatives of that bundle Φ. The aim of the present chapter is an investigation of flat complex vector bundles, and of their relationships with complex analytic vector bundles.

The case of flat complex line bundles (of flat vector bundles of rank 1, in other words) was discussed in last year's lectures, in connection with the classification of complex analytic line bundles; recall in particular §8. In that discussion, one began with the exact sequence of sheaves of abelian groups

$$(2) \qquad 0 \longrightarrow \mathbf{C}^* \xrightarrow{i} \mathcal{O}^* \xrightarrow{d\ell} \mathcal{O}^{1,0} \longrightarrow 0 ,$$

where i is again the inclusion mapping and $d\ell$ is the mapping defined by $d\ell(f) = \frac{1}{2\pi i} d \log f$, for any germ $f \in \mathcal{O}^*$. The exact cohomology sequence associated to (2) contains the segment

$$(3) \quad \Gamma(M, \mathcal{O}^{1,0}) \xrightarrow{\delta^*} H^1(M, \mathbf{C}^*) \xrightarrow{i^*} H^1(M, \mathcal{O}^*) \xrightarrow{d\ell^*} H^1(M, \mathcal{O}^{1,0}) \ ;$$

and by the Serre duality theorem, $H^1(M, \mathcal{O}^{1,0}) \cong \Gamma(M, \mathcal{O}) = \mathbb{C}$. As noted in last year's lectures (Lemma 19), the image $d\ell^*(\xi)$ of a complex line bundle $\xi \in H^1(M, \mathcal{O}^*)$ is essentially the Chern class of ξ ; hence <u>a line bundle ξ has a flat representative if and only if</u> $c(\xi) = 0$. Further, if $c(\xi) = 0$, the set of all flat representatives of ξ is the coset $\xi + \delta^*\Gamma(M, \mathcal{O}^{1,0})$, and the space of cosets is the Picard variety of M . The discussion of flat complex vector bundles of arbitrary rank is roughly parallel to the discussion of flat complex line bundles, as reviewed above; the non-abelian character of general vector bundles is rather a complicating factor, however. (For other treatments of this topic, see for instance A. Weil, "Généralization des fonctions abéliennes," J. Math. Pures Appl. 17(1938), 47-87; and M. F. Atiyah, "Complex analytic connections in fibre bundles," Trans. Amer. Math. Soc. 85 (1957), 181-207.)

One preliminary construction should first be discussed. Suppose that μ: $GL(m, \mathbb{C}) \longrightarrow GL(n, \mathbb{C})$ is a complex analytic representation; that is, μ is a complex analytic mapping between these two complex manifolds, and μ is a homomorphism of the group structures. It is obvious that μ induces a sheaf homomorphism

$\mu: \mathcal{GL}(m, \mathcal{O}) \longrightarrow \mathcal{GL}(n, \mathcal{O})$; and this in turn leads to a mapping between cohomology sets, of the form

$$\mu: H^1(M, \mathcal{GL}(m, \mathcal{O})) \longrightarrow H^1(M, \mathcal{GL}(n, \mathcal{O})) .$$

Thus to any complex analytic vector bundle $\Phi \in H^1(M, \mathcal{GL}(m, \mathcal{O}))$ there corresponds a complex analytic vector bundle $\mu\Phi \in H^1(M, \mathcal{GL}(n, \mathcal{O}))$. This construction will arise in particular for the adjoint representation $Ad: GL(m,\mathbb{C}) \longrightarrow GL(m^2,\mathbb{C})$, which is defined as follows. For any matrix $A \in GL(m,\mathbb{C})$, $Ad(A)$ is the linear transformation on the complex vector space $\mathbb{C}^{m \times m} = \mathbb{C}^{m^2}$ of $m \times m$ complex matrices, which associates to a matrix $X \in \mathbb{C}^{m \times m}$ the matrix $Ad(A)\cdot X \in \mathbb{C}^{m \times m}$ defined by

$$Ad(A)\cdot X = AXA^{-1} .$$

Thus there arises the cohomology mapping

$$Ad: H^1(M, \mathcal{GL}(m, \mathcal{O})) \longrightarrow H^1(M, \mathcal{GL}(m^2, \mathcal{O})) .$$

Now, to obtain the analogue of the exact sequence (2) for treating vector bundles of arbitrary rank, consider the differential operator D defined as follows. For any matrix $F \in GL(m, \mathcal{O}_U)$ over an open subset $U \subset M$, let $DF = F^{-1}dF$; thus DF is an $m \times m$ matrix of complex analytic differential forms of type $(1,0)$ over the set U , which will be written $DF \in (\mathcal{O}_U^{(1,0)})^{m \times m} = \mathcal{O}_U^{(1,0)} \otimes \mathcal{O}_U^{m \times m}$. The mapping D then leads to a sheaf mapping $D: \mathcal{GL}(m, \mathcal{O}) \longrightarrow (\mathcal{O}^{(1,0)})^{m \times m} = \mathcal{O}^{(1,0)} \otimes \mathcal{O}^{m \times m}$; and the kernel of the mapping D is the subsheaf $GL(m,\mathbb{C}) \subset \mathcal{GL}(m, \mathcal{O})$.

Lemma 17. Over any Riemann surface M, there is a twisted exact sequence of sheaves of groups of the form

$$(4) \quad 0 \longrightarrow GL(m,\mathbf{C}) \xrightarrow{i} \mathcal{GL}(m,\mathcal{O}) \xrightarrow{D} \mathcal{O}^{(1,0)} \otimes \mathcal{O}^{m\times m} \longrightarrow 0 \ ;$$

the inclusion mapping i is a homomorphism of sheaves of groups, while the sheaf mapping D satisfies

$$(5) \qquad D(FG) = \mathrm{Ad}(G^{-1})\cdot DF + DG \ .$$

Remark: To say that (4) is a twisted exact sequence of sheaves of (non-abelian) groups just means that i and D are sheaf mappings, with i a homomorphism and D satisfying a relation of the form (5), and that at each stage the kernel from the right is the image from the left, as usual.

Proof. It is clear that i is an isomorphism, with image precisely the subsheaf of $\mathcal{GL}(m,\mathcal{O})$ consisting of germs $F \in \mathcal{GL}(m,\mathcal{O})$ such that $DF = 0$. Moreover, if $\Lambda \in \mathcal{O}^{(1,0)}_p \otimes \mathcal{O}^{m\times m}_p = (\mathcal{O}^{(1,0)})^{m\times m}_p$, there is a germ $F \in \mathcal{O}^{m\times m}$ satisfying the differential equation $dF = F\Lambda$ and the initial condition $F(p) = I$; but then $F \in \mathcal{GL}(m,\mathcal{O})$ and $DF = \Lambda$, so the mapping D is onto. Finally, whenever $F, G \in \mathcal{GL}(m,\mathcal{O})_p$, it follows that $D(FG) = (FG)^{-1}d(FG) = G^{-1}F^{-1}(dFG + FdG) = G^{-1}DFG + DG = \mathrm{Ad}(G^{-1})\cdot DF + DG$ as desired, to conclude the proof.

Theorem 15. Let M be an arbitrary Riemann surface. To every complex analytic vector bundle $\Phi \in H^1(M, \mathcal{GL}(m,\mathcal{O}))$ there is associated a cohomology class $D^*\Phi \in H^1(M, \mathcal{O}(\kappa \otimes \mathrm{Ad}\Phi))$, where

$\kappa \in H^1(M, \mathcal{O}^*)$ is the canonical bundle; and Φ has a flat representative if and only if $D^*\Phi = 0$.

Proof. Suppose that $(\Phi_{\alpha\beta}) \in Z^1(\mathfrak{U}, \mathcal{GL}(m, \mathcal{O}))$ is a cocycle representing the vector bundle Φ , in terms of an open covering $\mathfrak{U} = \{U_\alpha\}$ of the Riemann surface M ; thus the matrices $\Phi_{\alpha\beta} \in GL(m, \mathcal{O}_{U_\alpha \cap U_\beta})$ satisfy $\Phi_{\alpha\beta}(p)\cdot\Phi_{\beta\gamma}(p) = \Phi_{\alpha\gamma}(p)$ whenever $p \in U_\alpha \cap U_\beta \cap U_\gamma$. Applying the differential operator D , it follows from (5) that

$$D\Phi_{\alpha\gamma}(p) = \mathrm{Ad}(\Phi_{\beta\gamma}(p)^{-1})\cdot D\Phi_{\alpha\beta}(p) + D\Phi_{\beta\gamma}(p) \quad \text{whenever} \quad p \in U_\alpha \cap U_\beta \cap U_\gamma ;$$

but this is just the condition that the elements $D\Phi_{\alpha\beta}$ form a one-cocycle $D\Phi_{\alpha\beta} \in Z^1(\mathfrak{U}, \mathcal{O}(\kappa \otimes \mathrm{Ad}\Phi))$, since $\mathcal{O}(\kappa \otimes \mathrm{Ad}\Phi) = \mathcal{O}^{(1,0)} \otimes \mathcal{O}(\mathrm{Ad}\Phi)$. (See the discussion in Appendix 1, for the cocycle formalism.) The cocycle $(\Phi_{\alpha\beta})$ is equivalent to a flat cocycle precisely when there is a zero-cochain $(F_\alpha) \in C^0(\mathfrak{U}, \mathcal{GL}(m, \mathcal{O}))$ such that $\Phi'_{\alpha\beta} = F_\alpha\Phi_{\alpha\beta}F_\beta^{-1}$ is flat, hence such that $D\Phi'_{\alpha\beta} = 0$. Applying (5) again, this condition is just that

$$\begin{aligned} 0 &= D(F_\alpha\Phi_{\alpha\beta}F_\beta^{-1}) \\ &= \mathrm{Ad}(F_\beta\Phi_{\alpha\beta}^{-1})\cdot DF_\alpha + \mathrm{Ad}(F_\beta)\cdot D\Phi_{\alpha\beta} + D(F_\beta^{-1}) \\ &= \mathrm{Ad}(F_\beta)\cdot[\mathrm{Ad}(\Phi_{\alpha\beta}^{-1})\cdot DF_\alpha + D\Phi_{\alpha\beta} - DF_\beta] . \end{aligned}$$

Putting $\lambda_\alpha = DF_\alpha$, this condition can be rewritten

$$D\Phi_{\alpha\beta} = \lambda_\beta - \mathrm{Ad}(\Phi_{\alpha\beta}^{-1})\cdot\lambda_\alpha ;$$

but this is equivalent to the assertion that the cocycle $D\Phi_{\alpha\beta}$ is the coboundary of the one-cochain $(\lambda_\alpha) \in C^0(\mathfrak{U}, \mathcal{O}(\kappa \otimes \mathrm{Ad}\Phi))$, where $\lambda_\alpha = DF_\alpha$ for some functions $F_\alpha \in GL(m, \mathcal{O}_{U_\alpha})$. Since any matrices λ_α can be written $\lambda_\alpha = DF_\alpha$, perhaps after passing to a refinement of the covering $\mathfrak{U}$, it follows that Φ has a flat representative precisely when $D\Phi_{\alpha\beta} = 0$ in $H^1(M, \mathcal{O}(\kappa \otimes \text{Ad}\,\Phi))$. Finally, it should be demonstrated that the construction is canonical, in the sense that the cohomology class associated to the bundle Φ is independent of the choice of representative cocycle $(\Phi_{\alpha\beta})$ of the bundle; this is a straightforward but uninteresting calculation, which will be left to the reader. That concludes the proof of the theorem.

(b) Recalling the notation introduced earlier (see p.69), let Φ^* denote the dual vector bundle to a vector bundle Φ; thus Φ^* is defined by the cocycle $({}^t\Phi_{\alpha\beta}^{-1})$, whenever $(\Phi_{\alpha\beta})$ is a cocycle defining Φ. It is then evident that $\text{Ad}\,\Phi^*$ is the dual vector bundle to the vector bundle $\text{Ad}\,\Phi$. Now from the Serre duality theorem for complex analytic vector bundles (Theorem 12), it follows that the cohomology groups $H^1(M, \mathcal{O}(\kappa \otimes \text{Ad}\,\Phi))$ and $H^0(M, \mathcal{O}(\text{Ad}\,\Phi^*))$ are canonically dual to one another; thus every cohomology class σ determines a linear functional (also denoted by σ) on the vector space $\Gamma(M, \mathcal{O}(\text{Ad}\,\Phi^*))$, and $\sigma = 0$ if and only if this associated linear functional is zero. It is of interest to have a more explicit form for this linear functional associated to a cohomology class σ. Recall that the Serre duality can be described explicitly as follows. Let $\mathfrak{U} = \{U_\alpha\}$ be an open covering of the Riemann surface M so that the vector bundle Φ is described by a cocycle $(\Phi_{\alpha\beta}) \in Z^1(\mathfrak{U}, \mathcal{L}(m, \mathcal{O}))$; and let $(\sigma_{\alpha\beta}) \in Z^1(\mathfrak{U}, \mathcal{O}(\kappa \otimes \mathrm{Ad}\,\Phi)) = Z^1(\mathfrak{U}, \mathcal{O}^{1,0}(\mathrm{Ad}\,\Phi))$ be a cocycle representing the cohomology class σ. The elements $\sigma_{\alpha\beta}$ can be viewed as column vectors of holomorphic differential forms of type $(1,0)$ associated to the various intersections $U_\alpha \cap U_\beta$, such that $\sigma_{\alpha\gamma} = \mathrm{Ad}(\Phi_{\beta\gamma}^{-1})\cdot\sigma_{\alpha\beta} + \sigma_{\beta\gamma}$ over $U_\alpha \cap U_\beta \cap U_\gamma$. Let $(\lambda_\alpha) \in C^0(\mathfrak{U}, \mathcal{E}^{1,0}(\mathrm{Ad}\,\Phi)) = C^0(\mathfrak{U}, \mathcal{E}(\kappa \otimes \mathrm{Ad}\,\Phi))$ be a zero-cochain with coboundary $(\sigma_{\alpha\beta})$; that is, let λ_α be vectors of C^∞ differential forms of type $(1,0)$ in the various sets U_α such that $\sigma_{\alpha\beta} = \lambda_\beta - \mathrm{Ad}(\Phi_{\alpha\beta}^{-1})\cdot\lambda_\alpha$ in $U_\alpha \cap U_\beta$. Since $\sigma_{\alpha\beta}$ are holomorphic, $\bar\partial\sigma_{\alpha\beta} = 0$; thus $\bar\partial\lambda_\alpha$ are column vectors of C^∞ differential forms of type $(1,1)$ such that $\bar\partial\lambda_\alpha = \mathrm{Ad}(\Phi_{\alpha\beta}^{-1})\cdot\bar\partial\lambda_\beta$ in $U_\alpha \cap U_\beta$. Now for any section $T = (T_\alpha) \in \Gamma(M, \mathcal{O}(\mathrm{Ad}\,\Phi^*))$, viewing T_α as column vectors of holomorphic functions in the various sets U_α such that $T_\alpha = \mathrm{Ad}(\Phi_{\alpha\beta}^*)T_\beta$ is $U_\alpha \cap U_\beta$, it follows that ${}^tT_\alpha\cdot\bar\partial\lambda_\alpha$ are scalar differential forms of type $(1,1)$ in the various sets U_α, and that ${}^tT_\alpha\cdot\bar\partial\lambda_\alpha = {}^tT_\beta\cdot\bar\partial\lambda_\beta$ in $U_\alpha \cap U_\beta$. Then the linear functional associated to σ is given explicitly by

$$\sigma(T) = \frac{1}{2\pi i}\int_M {}^tT_\alpha\cdot\bar\partial\lambda_\alpha \,.$$

Actually, it is more convenient to view the various elements λ_α and T_α as $m \times m$ matrices rather than as column vectors of length m^2, since the adjoint representation is then easier to describe; and with this convention in mind, the linear functional associated to $\sigma \in H^1(M, \mathcal{O}(\kappa \otimes \mathrm{Ad}\,\Phi))$ takes the value on the element $T \in \Gamma(M, \mathcal{O}(\mathrm{Ad}\,\Phi^*))$ given explicitly by

(6) $$\sigma(T) = \frac{1}{2\pi i}\int_M \mathrm{tr}({}^tT_\alpha\cdot\bar\partial\lambda_\alpha) \,,$$

where tr denotes the trace of a matrix.

The cohomology class $\sigma = D^{*}\Phi \in H^{1}(M, \mathcal{O}(\kappa \otimes \text{Ad } \Phi))$ is of course of particular interest, and it can be described explicitly as a linear functional on $\Gamma(M, \mathcal{O}(\text{Ad } \Phi^{*}))$ as follows. Select matrices $F_{\alpha} \in GL(m, \mathcal{M}_{U_\alpha})$ such that $\Phi_{\alpha\beta} = F_{\alpha}F_{\beta}^{-1}$ over $U_{\alpha} \cap U_{\beta}$; this is always possible, since all meromorphic vector bundles are trivial (as pointed out on p.43). Furthermore, assume that the covering $\mathcal{U} = \{U_{\alpha}\}$ is so chosen that the singularities of the matrices F_{α} (the poles of any entry or the points where the matrix has zero determinant) lie in disjoint open sets U_{j}, and do not lie in any intersection $U_{j} \cap U_{\alpha}$. Now from (5) it follows that

$$\begin{aligned}\sigma_{\alpha\beta} &= D\Phi_{\alpha\beta} = D(F_{\alpha}F_{\beta}^{-1}) \\ &= \text{Ad}(F_{\beta})\cdot DF_{\alpha} + D(F_{\beta}^{-1}) \\ &= \text{Ad}(\Phi_{\alpha\beta}^{-1})\cdot\text{Ad}(F_{\alpha})\cdot DF_{\alpha} - \text{Ad}(F_{\beta})\cdot DF_{\beta}\ ;\end{aligned}$$

thus

$$\tilde{\lambda}_{\alpha} = -\text{Ad}(F_{\alpha})\cdot DF_{\alpha}$$

are $m \times m$ matrices of meromorphic differential forms of type $(1,0)$ in the sets U_{α}, such that $\sigma_{\alpha\beta} = \tilde{\lambda}_{\beta} - \text{Ad}(\Phi_{\alpha\beta}^{-1})\tilde{\lambda}_{\alpha}$ in $U_{\alpha} \cap U_{\beta}$. For each set U_{α} select a C^{∞} function r_{α} such that $r_{\alpha} \equiv 1$ on each intersection $U_{\alpha} \cap U_{\beta}$, that r_{j} vanishes identically in an open neighborhood of each singularity of the meromorphic differential form $\tilde{\lambda}_{j}$ in U_{j}, and that $r_{\alpha} \equiv 1$ if $\tilde{\lambda}_{\alpha}$ is non-singular in U_{α}; this is always possible, in view of the special form of the covering $\mathcal{U}$. Then the differential forms $\lambda_{\alpha} = r_{\alpha}\tilde{\lambda}_{\alpha}$ are C^{∞} in each set U_{α}, and also satisfy the relation $\sigma_{\alpha\beta} = \lambda_{\beta} - \text{Ad}(\Phi_{\alpha\beta}^{-1})\lambda_{\alpha}$

in $U_\alpha \cap U_\beta$. Thus by (6) the linear functional associated to $\sigma = D^*\Phi$ is given by

$$D^*\Phi(T) = \frac{1}{2\pi i} \int_M \operatorname{tr}({}^tT_\alpha \cdot \bar{\partial}\lambda_\alpha)$$

for any element $T = (T_\alpha) \in \Gamma(M, \mathcal{O}(\mathrm{Ad}\ \Phi^*))$. Now since λ_α is holomorphic except for those sets U_j containing the singularities of $\tilde{\lambda}_j$, and the elements T_α are all holomorphic, this expression can be rewritten

$$\begin{aligned} 2\pi i D^*\Phi(T) &= \sum_j \int_{U_j} \bar{\partial}\operatorname{tr}({}^tT_j\lambda_j) = \sum_j \int_{U_j} d\operatorname{tr}({}^tT_j\lambda_j) \\ &= \sum_j \int_{\partial U_j} \operatorname{tr}({}^tT_j\lambda_j) = \sum_j \int_{\partial U_j} \operatorname{tr}({}^tT_j\tilde{\lambda}_j) \\ &= \sum_j 2\pi i\, \mathcal{R}\operatorname{tr}({}^tT_j\tilde{\lambda}_j)\ , \end{aligned}$$

where $\mathcal{R}$ denotes the total residue of the meromorphic differential form at all its singularities; therefore the linear functional associated to the cohomology class $D^*\Phi$ has the value

$$(7)\qquad \begin{aligned} D^*\Phi(T) &= -\mathcal{R}\operatorname{tr}({}^tT_\alpha\ \mathrm{Ad}(F_\alpha)\cdot DF_\alpha) \\ &= -\mathcal{R}\operatorname{tr}({}^tT_\alpha\ dF_\alpha\ F_\alpha^{-1})\ , \end{aligned}$$

for any section $T = (T_\alpha) \in \Gamma(M, \mathcal{O}(\mathrm{Ad}\ \Phi^*))$. With this observation, Theorem 15 can be given the following restatement.

<u>Corollary 1.</u> Let $\Phi \in H^1(M, \mathcal{GL}(m, \mathcal{O}))$ be a complex analytic vector bundle over the compact Riemann surface M; let $(\Phi_{\alpha\beta}) \in Z^1(\mathfrak{U}, \mathcal{GL}(m, \mathcal{O}))$ be a representative cocycle for Φ, and let $(F_\alpha) \in C^0(\mathfrak{U}, \mathcal{GL}(m, \mathcal{M}))$ be meromorphic matrices such that $F_\alpha = \Phi_{\alpha\beta}F_\beta$ in $U_\alpha \cap U_\beta$. Then Φ has flat representatives if and only if

$$\mathcal{R}\, \mathrm{tr}({}^t T_\alpha\, dF_\alpha\, F_\alpha^{-1}) = 0$$

for all sections $T = (T_\alpha) \in \Gamma(M, \mathcal{O}(\mathrm{Ad}\ \Phi^*))$, where $\mathcal{R}$ denotes the total residue of the differential form $\mathrm{tr}({}^t T_\alpha\, dF_\alpha\, F_\alpha^{-1}) \in \Gamma(M, \mathcal{M}^{1,0})$.

The condition for a bundle to have flat representatives, as restated in Corollary 1, is actually quite useful, after a few preliminary observations about the space $\Gamma(M, \mathcal{O}(\mathrm{Ad}\ \Phi^*))$. For any complex analytic vector bundle Φ , an <u>endomorphism</u> of Φ is defined to be a sheaf homomorphism $T\colon \mathcal{O}(\Phi) \longrightarrow \mathcal{O}(\Phi)$. Let $\mathcal{U} = \{U_\alpha\}$ be an open covering of the Riemann surface M , such that the vector bundle Φ has a representative cocycle $(\Phi_{\alpha\beta}) \in Z^1(\mathcal{U}, \mathcal{GL}(m, \mathcal{O}))$. Then the restriction of the sheaf $\mathcal{O}(\Phi)$ to each set U_α has an isomorphic representation as a free sheaf, $\mathcal{O}(\Phi)|U_\alpha \cong \mathcal{O}^m|U_\alpha$; so that an endomorphism T determines homomorphisms $T_\alpha\colon \mathcal{O}^m|U_\alpha \longrightarrow \mathcal{O}^m|U_\alpha$ of free sheaves and these are described by matrices $T_\alpha \in \Gamma(U_\alpha, \mathcal{O}^{m\times m})$. It is readily seen that these matrices satisfy the conditions

$$(8) \qquad T_\alpha(p)\Phi_{\alpha\beta}(p) = \Phi_{\alpha\beta}(p)T_\beta(p) \quad \text{whenever} \quad p \in U_\alpha \cap U_\beta\ ;$$

and conversely, any set of matrices $(T_\alpha) \in C^0(\mathcal{U}, \mathcal{O}^{m\times m})$ satisfying (8) determines an endomorphism of the vector bundle Φ represented by the cocycle $(\Phi_{\alpha\beta})$. It is clear that the set of endomorphisms of Φ , which will be denoted by $\mathrm{End}\ \Phi$, has the structure of an algebra over the complex numbers; if $S = (S_\alpha)$ and $T = (T_\alpha)$ are endomorphisms, and a, b are complex numbers, let $aS + bT \in \mathrm{End}\ \Phi$ be the endomorphism represented by the matrices

$(aS_\alpha + bT_\alpha)$, and let $ST \in \text{End}\,\Phi$ be the endomorphism represented by the matrices $(S_\alpha T_\alpha)$. Now consider any section $R = (R_\alpha) \in \Gamma(M, \mathcal{O}(\text{Ad}\,\Phi^*))$; the elements R_α can be viewed as holomorphic $m \times m$ matrices defined over the sets U_α , and these matrices satisfy $R_\alpha = \text{Ad}(\Phi^*_{\alpha\beta})\cdot R_\beta = \Phi^*_{\alpha\beta}\, R_\beta\, (\Phi^*_{\alpha\beta})^{-1} = {}^t\Phi^{-1}_{\alpha\beta}\, R_\beta\, {}^t\Phi_{\alpha\beta}$ in $U_\alpha \cap U_\beta$. Note then that $T_\alpha = {}^tR_\alpha$ satisfies $T_\alpha = \Phi_{\alpha\beta} T_\beta \Phi^{-1}_{\alpha\beta}$, so that $T = (T_\alpha) \in \text{End}\,\Phi$; it is thus clear that there is a natural one-to-one correspondence between the sets $\Gamma(M, \mathcal{O}(\text{Ad}\,\Phi^*))$ and $\text{End}\,\Phi$. With this observation, Corollary 1 to Theorem 15 can be restated as follows.

Corollary 2. Let $\Phi \in H^1(M, \mathcal{GL}(m, \mathcal{O}))$ be a complex analytic vector bundle over the compact Riemann surface M ; let $(\Phi_{\alpha\beta}) \in Z^1(\mathfrak{U}, \mathcal{GL}(m, \mathcal{O}))$ be a representative cocycle for Φ , and let $(F_\alpha) \in C^0(\mathfrak{U}, \mathcal{GL}(m, \mathcal{M}))$ be meromorphic matrices such that $F_\alpha = \Phi_{\alpha\beta} F_\beta$ in $U_\alpha \cap U_\beta$. Then Φ has a flat representative if and only if

$$\mathcal{R}\ \text{tr}(T_\alpha\, dF_\alpha\, F_\alpha^{-1}) = 0$$

for all endomorphisms $T = (T_\alpha) \in \text{End}\,\Phi$.

Some further consequences of Theorem 15 now follow rather readily, upon looking more closely at the algebra $\text{End}\,\Phi$ of endomorphisms of the bundle Φ .

Corollary 3. Let $\Phi_i \in H^1(M, \mathcal{GL}(m_i, \mathcal{O}))$, $i = 1,\ldots,r$, be complex analytic vector bundles over a compact Riemann surface M , and let $\Phi = \Phi_1 \oplus \ldots \oplus \Phi_r$. Then Φ has flat representatives if and only if all the bundles Φ_i have flat representatives.

Proof. In terms of a suitable coordinate covering $\mathfrak{U} = \{U_\alpha\}$ the bundle Φ can be represented by a cocycle $\Phi_{\alpha\beta} = \Phi_{1\alpha\beta} \oplus \ldots \oplus \Phi_{r\alpha\beta}$, where $\Phi_{i\alpha\beta}$ are cocycles representing Φ_i ; and selecting matrices $F_{i\alpha} \in C^0(\mathfrak{U}, \mathcal{GL}(m_i, \mathcal{M}))$ satisfying $F_{i\alpha} = \Phi_{i\alpha\beta} F_{i\beta}$, it follows that the matrices $F_\alpha = F_{1\alpha} \oplus \ldots \oplus F_{r\alpha}$ satisfy $F_\alpha = \Phi_{\alpha\beta} F_\beta$. Now if $T = (T_\alpha)$ is any endomorphism of the bundle Φ, the matrices T_α can be decomposed into r^2 blocks $T_\alpha = T_{\alpha ij}$, $1 \leq i, j \leq r$, corresponding to the decomposition of Φ ; and $T_{ii} = (T_{ii\alpha})$ will be an endomorphism of Φ_i. Conversely, if $T_{ii} = (T_{ii\alpha})$ are arbitrary endomorphisms of Φ_i, $T = T_{11} \oplus \ldots \oplus T_{rr} = (T_{11\alpha} \oplus \ldots \oplus T_{rr\alpha})$ will be an endomorphism of Φ. Note that

$$\operatorname{tr}(T_\alpha \, dF_\alpha \, F_\alpha^{-1}) = \operatorname{tr}(T_\alpha \, (dF_{1\alpha} \, F_{1\alpha}^{-1} \oplus \ldots \oplus dF_{r\alpha} \, F_{r\alpha}^{-1}))$$

$$= \sum_{i=1}^{r} \operatorname{tr}(T_{ii\alpha} \, dF_{i\alpha} \, F_{i\alpha}^{-1}) .$$

But then the desired result follows immediately from Corollary 2, and the proof is thereby concluded.

Now consider any endomorphism $T = (T_\alpha) \in \operatorname{End} \Phi$; here T_α are holomorphic $m \times m$ matrices in the various sets U_α, and $T_\alpha(p) = \Phi_{\alpha\beta}(p) T_\beta(p) \Phi_{\alpha\beta}(p)^{-1}$ whenever $p \in U_\alpha \cap U_\beta$. The characteristic polynomial of the matrix T_α thus coincides with the characteristic polynomial of the matrix T_β in $U_\alpha \cap U_\beta$; so that the characteristic polynomial of the endomorphism T is a well-defined polynomial of degree m, with coefficients which are holomorphic functions on the Riemann surface M. <u>If</u> M <u>is compact,</u> the coefficients must be constant; but then the roots of the polynomial, which are the eigenvalues of the matrices T_α, are also constants. The identity endomorphism $I \in \text{End}\,\Phi$, where the matrices I_α are all identity matrices; all eigenvalues of the identity are 1. An endomorphism $T \in \text{End}\,\Phi$ is called nilpotent if $T^r = 0$ for some integer r, where 0 is the zero endomorphism; it is obvious that the nilpotent endomorphisms are precisely those for which all the eigenvalues are zero.

Lemma 18. Let $\Phi \in H^1(M, \mathfrak{b}\mathfrak{L}(m, \mathcal{O}))$ be an indecomposable complex analytic vector bundle over a compact Riemann surface M. Then any endomorphism $T = (T_\alpha) \in \text{End}\,\Phi$ can be written uniquely as a sum $T = cI + N$, where $c \in \mathbb{C}$, I is the identity endomorphism, and N is a nilpotent endomorphism.

Proof. Note that an endomorphism of the form $T = cI + N$ has all eigenvalues equal to c; and conversely, if all eigenvalues of T are equal to c, the endomorphism $N = T - cI$ has all eigenvalues equal to 0, hence N is nilpotent. Therefore the desired lemma is equivalent to the assertion that, with the same hypotheses, an endomorphism $T \in \text{End}\,\Phi$ has all eigenvalues coinciding.

To prove that this is so, suppose contrariwise that an endomorphism T has at least two distinct eigenvalues. Then, in terms of a suitable coordinate covering $\mathfrak{U} = \{U_\alpha\}$ and a suitable cocycle $(\Phi_{\alpha\beta}) \in Z^1(\mathfrak{U}, \mathfrak{b}\mathfrak{L}(m, \mathcal{O}))$ representing the bundle Φ, the endomorphism $T = (T_\alpha) \in \text{End}\,\Phi$ will be represented by matrices of the form $T_\alpha = T_{1\alpha} \oplus T_{2\alpha}$, where $T_{1\alpha}$ and $T_{2\alpha}$ have different eigenvalues. Now decomposing the matrices $\Phi_{\alpha\beta}$ into four corresponding blocks, the condition that T be an endomorphism of Φ

is just that

$$\begin{pmatrix} T_{1\alpha} & 0 \\ 0 & T_{2\alpha} \end{pmatrix} \begin{pmatrix} \Phi_{11\alpha\beta} & \Phi_{12\alpha\beta} \\ \Phi_{21\alpha\beta} & \Phi_{22\alpha\beta} \end{pmatrix} = \begin{pmatrix} \Phi_{11\alpha\beta} & \Phi_{12\alpha\beta} \\ \Phi_{21\alpha\beta} & \Phi_{22\alpha\beta} \end{pmatrix} \begin{pmatrix} T_{1\beta} & 0 \\ 0 & T_{2\beta} \end{pmatrix}$$

in each intersection $U_\alpha \cap U_\beta$. But this implies that $T_{1\alpha}\Phi_{12\alpha\beta} = \Phi_{12\alpha\beta}T_{2\beta}$ and $T_{2\alpha}\Phi_{21\alpha\beta} = \Phi_{21\alpha\beta}T_{1\beta}$. Since T_1 and T_2 have altogether different eigenvalues, it follows that $\Phi_{12} = 0$ and $\Phi_{21} = 0$, hence that the bundle Φ is decomposable; this is a contradiction, so the proof is thereby concluded.

<u>Lemma 19.</u> Let $\Phi \in H^1(M, \mathcal{GL}(m, \mathcal{O}))$ be a complex analytic vector bundle over an arbitrary Riemann surface M, and let $T \in \operatorname{End}\Phi$ be an endomorphism of the bundle Φ. Then there are representatives $\Phi = (\Phi_{\alpha\beta}) \in Z^1(\mathcal{U}, \mathcal{GL}(m, \mathcal{O}))$ and $T = (T_\alpha)$, such that the matrices $\Phi_{\alpha\beta}$ and T_α are all in upper triangular form.

Proof. Let $\varphi \subset \Phi$ be a complex analytic line bundle contained as a subbundle $\varphi \subset \Phi$ and with $c(\varphi) = \operatorname{div}\Phi$. The set $\Gamma(M, \mathcal{O}(\varphi^{-1} \otimes \Phi))$ is then a non-trivial finite-dimensional complex vector space, and all the non-trivial sections are non-singular holomorphic vector-valued functions, (that is, have trivial divisors). It is clear that T also determines an endomorphism of $\varphi^{-1} \otimes \Phi$; so T acts as a linear transformation on the vector space $\Gamma(M, \mathcal{O}(\varphi^{-1} \otimes \Phi))$. Let $F \in \Gamma(M, \mathcal{O}(\varphi^{-1} \otimes \Phi))$ be a non-trivial section which is an eigenvector of the operator T; and choose representatives $\Phi = (\Phi_{\alpha\beta}) \in Z^1(\mathcal{U}, \mathcal{GL}(m, \mathcal{O}))$ and $T = (T_\alpha) \in C^0(\mathcal{U}, \mathcal{O}^{m \times m})$ for a suitable open covering $\mathcal{U}$, such that the section F is represented by $F_\alpha = E_1$ (where as before, $E_1 = \begin{pmatrix} 1 \\ 0 \\ \cdots \\ 0 \end{pmatrix}$). Now since $E_1 = F_\alpha = \varphi_{\alpha\beta}^{-1}\Phi_{\alpha\beta}F_\beta = \varphi_{\alpha\beta}^{-1}\Phi_{\alpha\beta}E_1$, it follows that the matrices $\Phi_{\alpha\beta}$ have the form

$$\Phi_{\alpha\beta} = \begin{pmatrix} \varphi_{\alpha\beta} & * & \cdots & * \\ 0 & * & \cdots & * \\ \cdots & \cdots & & \cdots \\ 0 & * & \cdots & * \end{pmatrix};$$

and since $T_\alpha E_1 = T_\alpha F_\alpha = \lambda F_\alpha = \lambda E_1$, where λ is the eigenvalue of T_α for the eigenvector F, it also follows that the matrices T_α have the form

$$T_\alpha = \begin{pmatrix} \lambda & * & \cdots & * \\ 0 & * & \cdots & * \\ \cdots & \cdots & & \cdots \\ 0 & * & \cdots & * \end{pmatrix}.$$

The desired result then follows from the obvious induction.

With these two observations made, the most important consequence of Theorem 15 follows quite readily.

<u>Theorem 16.</u> (Weil's theorem) Let M be a compact Riemann surface, and $\Phi = \Phi_1 \oplus \ldots \oplus \Phi_r$ be a complex analytic vector bundle over that Riemann surface, where the bundles Φ_i are indecomposable. Then Φ has flat representatives if and only if $c(\det \Phi_i) = 0$ for all i.

Proof. It follows immediately from Corollary 3 of Theorem 15 that it suffices to prove that an indecomposable vector bundle Φ has flat representatives if and only if $c(\det \Phi) = 0$. So suppose that $\Phi \in H^1(M, \mathcal{O}\mathcal{L}(m, \mathcal{O}))$ is indecomposable, and let $D^*\Phi$ be the cohomology class given in Theorem 15, viewing $D^*\Phi$ as a linear functional on the complex vector space $\operatorname{End}\Phi$ of endomorphisms of Φ. In view of Lemma 18, the functional $D^*\Phi$ is fully determined once its values on the identity endomorphism and on the nilpotent endomorphisms are determined. On the one hand, note that $D^*\Phi(I) = - \mathcal{R}\operatorname{tr}(dF_\alpha F_\alpha^{-1}) = - \mathcal{R}\, d \log \det F_\alpha$, recalling formula (7); for it is familiar - or an easy calculation if not familiar - that $\operatorname{tr}(dFF^{-1}) = d \log \det F$ for any non-singular matrix-valued function F. The functions $f_\alpha = \det F_\alpha$, using the notation in the discussion preceding formula (7), form a meromorphic section of the line bundle $\det \Phi$; and $\mathcal{R}\, d \log f_\alpha = \Sigma_{p \in M} \nu_p(f_\alpha)$. Therefore, by Theorem 11 of last year's lectures, $d \log f_\alpha = c(\det \Phi)$; so that

$$D^*\Phi(I) = - c(\det \Phi) . \tag{9}$$

On the other hand, for a nilpotent endomorphism T, apply Lemma 19 to choose representatives $\Phi = (\Phi_{\alpha\beta}) \in Z^1(\mathcal{U}, \mathcal{O}\mathcal{L}(m, \mathcal{O}))$ and $T = (T_\alpha)$ which are upper triangular matrices; and choose meromorphic non-singular matrices F_α also in upper triangular form and satisfying $F_\alpha = \Phi_{\alpha\beta}F_\beta$. Then since all the matrices T_α, dF_α, F_α^{-1} are in upper triangular form, the diagonal entries in $T_\alpha\, dF_\alpha\, F_\alpha^{-1}$ will be the products of the corresponding diagonal terms in the factors; but the diagonal entries of T_α are its eigenvalues, and are all zero since T_α is nilpotent. Thus

$$D^*\Phi(T) = 0 \quad \text{if} \quad T \in \operatorname{End}\Phi \quad \text{is nilpotent.} \tag{10}$$

Combining assertions (9) and (10), it follows that $D^*\Phi = 0$ if and only if $c(\det \Phi) = 0$; then by Theorem 15, it further follows that Φ has flat representatives if and only if $c(\det \Phi) = 0$, which suffices to conclude the proof.

(c) If Φ is a complex analytic vector bundle admitting flat representatives, there naturally arises the problem of describing all of the flat representatives of Φ . Note that from Theorem 16, if Φ has flat representatives, necessarily $c(\det \Phi) = 0$; so the line bundle $\det \Phi$ has a flat representative $\varphi \in H^1(M, \mathbf{C}^*)$. Then tensoring by φ establishes a one-to-one correspondence between all the flat representatives of $\varphi^{-1} \otimes \Phi$ and all those of Φ ; so for the further discussion, it suffices to restrict consideration to the set of vector bundles

(11) $$\mathcal{V}_m = \mathcal{V}_m(1) = \{\Phi \in H^1(M, \mathfrak{b}\mathfrak{X}(m, \mathcal{O}))|\det \Phi = 1\} .$$

The problem is to describe the set of flat representatives of any bundle $\Phi \in \mathcal{V}_m$.

For any bundle $\Phi \in \mathcal{V}_m$, select a representative cocycle $(\Phi_{\alpha\beta}) \in Z^1(\mathfrak{U}, \mathfrak{b}\mathfrak{X}(m, \mathcal{O}))$, and consider the one-cocycle $D\Phi = (D\Phi_{\alpha\beta}) \in Z^1(\mathfrak{U}, \mathcal{O}(\kappa \otimes \mathrm{Ad}\ \Phi))$ introduced in Theorem 15. A <u>complex analytic connection</u> for the bundle Φ in terms of the covering $\mathfrak{U}$ is a zero-cochain $\lambda = (\lambda_\alpha) \in C^0(\mathfrak{U}, \mathcal{O}(\kappa \otimes \mathrm{Ad}\ \Phi))$ such that $\delta\lambda = D\Phi$; such connections exist, in view of Theorem 15, precisely when the bundle Φ has flat representatives. The set of all such connections will be denoted by $\Lambda(\mathfrak{U}, \Phi)$. If $\lambda, \lambda' \in \Lambda(\mathfrak{U}, \Phi)$ are two such connections, the differences $\tau_\alpha = \lambda'_\alpha - \lambda_\alpha$ form a one-cochain $\tau \in C^0(\mathfrak{U}, \mathcal{O}(\kappa \otimes \mathrm{Ad}\ \Phi))$ such that $\delta\tau = \delta\lambda' - \delta\lambda = D\Phi - D\Phi = 0$;

that is to say, these elements determine a section $\tau \in \Gamma(M, \mathcal{O}(\kappa \otimes \text{Ad } \Phi))$. The set $\Lambda(\mathfrak{U}, \Phi) \subset C^0(\mathfrak{U}, \mathcal{O}(\kappa \otimes \text{Ad } \Phi))$ of all such connections is thus a coset of the subgroup $\Gamma(M, \mathcal{O}(\kappa \otimes \text{Ad } \Phi)) \subset C^0(\mathfrak{U}, \mathcal{O}(\kappa \otimes \text{Ad } \Phi))$ of sections (zero-cocyles) of the vector bundle $\kappa \otimes \text{Ad } \Phi$; so that, on a compact Riemann surface M, the set of connections $\Lambda(\mathfrak{U}, \Phi)$ is naturally a finite-dimensional linear space. (As to the terminology, the problem of describing the flat representatives of a complex analytic vector bundle Φ can be viewed as a problem of reducing the pseudogroup structure on the bundle space, viewed as a complex manifold. A general class of similar pseudogroup problems can be treated by basically the same formalism that is involved here; this includes problems leading to the usual affine and projective connections on manifolds, and connections in vector bundles, and the terminology has been adapted from the geometers. For a discussion of connections from this point of view, see R. C. Gunning, "Connections for a class of pseudogroup structure," Proc. Conf. on Complex Analysis (Minnesota, 1964), Springer-Verlag, 1965.)

If T is any automorphism of the vector bundle Φ, T induces in a natural manner an automorphism $\text{Ad } T$ of the vector bundle $\text{Ad } \Phi$, or of the vector bundle $\kappa \otimes \text{Ad } \Phi$. (For the condition that $T = (T_\alpha)$ be an automorphism is just that the matrices $T_\alpha \in GL(m, \mathcal{O}_{U_\alpha})$ satisfy $T_\alpha(p)\Phi_{\alpha\beta}(p) = \Phi_{\alpha\beta}(p)T_\beta(p)$ whenever $p \in U_\alpha \cap U_\beta$; and then it is evident that $\text{Ad } T_\alpha(p)\cdot\text{Ad } \Phi_{\alpha\beta}(p) = \text{Ad } \Phi_{\alpha\beta}(p)\cdot\text{Ad } T_\beta(p)$, or that the same relation holds upon tensoring with any line bundle.) The mapping $\text{Ad } T$ in turn induces isomorphisms $\mathrm{Ad}\,T: C^q(\mathfrak{U}, \mathcal{O}(\kappa \otimes \mathrm{Ad}\,\Phi)) \longrightarrow C^q(\mathfrak{U}, \mathcal{O}(\kappa \otimes \mathrm{Ad}\,\Phi))$ which commute with the coboundary mappings. It is of interest to see an explicit form for these mappings. In terms of the natural identifications, a zero-cochain $\lambda \in C^0(\mathfrak{U}, \mathcal{O}(\kappa \otimes \mathrm{Ad}\,\Phi))$ can be represented by holomorphic differential forms λ_α of type (1,0) in the various sets U_α, and $\mathrm{Ad}\,T\cdot\lambda$ will be represented by the differential forms $\mathrm{Ad}\,T_\alpha\cdot\lambda_\alpha$; and a one-cochain $\sigma \in C^1(\mathfrak{U}, \mathcal{O}(\kappa \otimes \mathrm{Ad}\,\Phi))$ can be represented by holomorphic differential forms $\sigma_{\alpha\beta}$ of type (1,0) in the various intersections $U_\alpha \cap U_\beta$, and $\mathrm{Ad}\,T\cdot\sigma$ will be represented by the differential forms $\mathrm{Ad}\,T_\beta\cdot\sigma_{\alpha\beta}$. (See also the discussion in Appendix 1.) It is of particular interest to consider the effect of the mapping $\mathrm{Ad}\,T$ on the cocycle $D\Phi = (D\Phi_{\alpha\beta})$. From the definition of an automorphism, $\Phi_{\alpha\beta} = T_\alpha\Phi_{\alpha\beta}T_\beta^{-1}$ in each intersection $U_\alpha \cap U_\beta$; applying the differential operator D, in view of (5) it follows that

$$\begin{aligned} D\Phi_{\alpha\beta} &= \mathrm{Ad}(T_\beta\Phi_{\alpha\beta}^{-1})\cdot DT_\alpha + \mathrm{Ad}(T_\beta)\cdot D\Phi_{\alpha\beta} + DT_\beta^{-1} \\ &= \mathrm{Ad}(\Phi_{\alpha\beta}^{-1}T_\alpha)\cdot DT_\alpha + \mathrm{Ad}(T_\beta)\cdot D\Phi_{\alpha\beta} - \mathrm{Ad}(T_\beta)\cdot DT_\beta\ . \end{aligned}$$

Setting $\tau_\alpha = \mathrm{Ad}(T_\alpha)\cdot DT_\alpha$, and viewing these elements as forming a cochain $\tau = (\tau_\alpha) \in C^0(\mathfrak{U}, \mathcal{O}(\kappa \otimes \mathrm{Ad}\,\Phi))$, the above condition can be rewritten

$$\mathrm{Ad}(T_\beta)\cdot D\Phi_{\alpha\beta} - D\Phi_{\alpha\beta} = \tau_\beta - \mathrm{Ad}(\Phi_{\alpha\beta}^{-1})\cdot\tau_\alpha\ ;$$

or, in other words,

$$\mathrm{Ad}(T)\cdot D\Phi - D\Phi = \delta\tau\ . \tag{12}$$

Thus the cocycles $D\Phi$ and $\mathrm{Ad}(T)\cdot D\Phi$ are cohomologous; indeed, formula (12) gives an explicit coboundary relating these two cocycles, in terms of the automorphism T. Now if $\lambda = (\lambda_\alpha) \in \Lambda(\mathfrak{U}, \Phi)$ is a connection, so that $D\Phi = \delta\lambda$, upon applying the operator $\mathrm{Ad}\, T$ it follows that $\mathrm{Ad}(T)\cdot D\Phi = \mathrm{Ad}(T)\cdot\delta\lambda = \delta\, \mathrm{Ad}(T)\cdot\lambda$; therefore by (12),

$$\begin{aligned} D\Phi &= \mathrm{Ad}(T)\cdot D\Phi - \delta\tau \\ &= \delta[\mathrm{Ad}(T)\cdot\lambda - \tau] , \end{aligned}$$

so that $\mathrm{Ad}(T)\cdot\lambda - \tau$ is also a connection for the bundle Φ. Thus to each automorphism $T \in \mathrm{Aut}\,\Phi$ there is associated a mapping

(13) $$\mathrm{Ad}^*(T) : \Lambda(\mathfrak{U}, \Phi) \longrightarrow \Lambda(\mathfrak{U}, \Phi)$$

defined by

(14) $$\mathrm{Ad}^*(T)\cdot\lambda = \mathrm{Ad}(T)\cdot\lambda - \tau ;$$

it is an easy matter to verify that the mapping (13) exhibits $\mathrm{Aut}\,\Phi$ as a group of transformations on $\Lambda(\mathfrak{U}, \Phi)$, but the details will be left to the reader. Two connections $\lambda, \lambda' \in \Lambda(\mathfrak{U}, \Phi)$ will be called <u>equivalent,</u> denoted by $\lambda \sim \lambda'$, if $\lambda' = \mathrm{Ad}^*(T)\cdot\lambda$ for some automorphism $T \in \mathrm{Aut}\,\Phi$; it is clear that this is an equivalence relation, indeed, is just the equivalence relation associated to the group action of $\mathrm{Aut}\,\Phi$ on the set $\Lambda(\mathfrak{U}, \Phi)$. The set of equivalence classes of connections will be denoted by $\Lambda^*(\mathfrak{U}, \Phi)$.

<u>Theorem 17.</u> Let M be an arbitrary Riemann surface, and $\Phi \in \mathcal{V}_m(1)$ be a complex analytic vector bundle over M represented by a cocycle $(\Phi_{\alpha\beta}) \in Z^1(\mathfrak{U}, \mathcal{GL}(m, \mathcal{O}))$ in terms of an open covering $\mathfrak{U} = \{U_\alpha\}$ of M ; and suppose the covering $\mathfrak{U}$ is such that all the sets U_α are contractible. Then there is a one-to-one correspondence between the set of flat representatives of Φ in the covering $\mathfrak{U}$, and the set $\Lambda^*(\mathfrak{U}, \Phi)$ of equivalence classes of connections for the bundle Φ.

Proof. If $\lambda = (\lambda_\alpha) \in \Lambda(\mathcal{U},\Phi)$ is any complex analytic connection, there are functions $F_\alpha \in GL(m, \mathcal{O}_{U_\alpha})$ such that $DF_\alpha = \lambda_\alpha$ in each set U_α ; that there exist local solutions of the equation $DF_\alpha = \lambda_\alpha$, or equivalently of the equation $dF_\alpha = F_\alpha\lambda_\alpha$, is familiar, and these yield global solutions in any simply-connected set U_α . The most general solution is readily seen to be of the form $C_\alpha F_\alpha$, for some matrix $C_\alpha \in GL(m,\mathbf{C})$; for if $DF_\alpha = DG_\alpha$, it follows from equation (5) that $D(F_\alpha G_\alpha^{-1}) = Ad(G_\alpha)\cdot DF_\alpha + DG_\alpha^{-1} = $ $= Ad(G_\alpha)\cdot[DF_\alpha - DG_\alpha] = 0$, hence $F_\alpha G_\alpha^{-1} = C_\alpha^{-1}$ is constant. Then as in the proof of Theorem 15, it follows that $\Phi'_{\alpha\beta} = F_\alpha \Phi_{\alpha\beta} F_\beta^{-1}$ represents the same vector bundle, and since it readily follows from (5) that $D\Phi'_{\alpha\beta} = 0$, this yields a flat representative. Of course, replacing F_α by $C_\alpha F_\alpha$ yields the same flat representative, as an element of $H^1(M,GL(m,\mathbf{C}))$. Thus each connection $\lambda \in \Lambda(\mathcal{U},\Phi)$ yields a well-defined flat representative of the bundle Φ ; and conversely, as in Theorem 15, all the flat representatives of Φ in terms of the covering $\mathcal{U}$ so arise.

Now two connections $\lambda = (\lambda_\alpha) = (DF_\alpha)$ and $\lambda' = (\lambda'_\alpha) = (DF'_\alpha)$ yield the same flat representative of the bundle Φ if and only if there are matrices $C_\alpha \in GL(m,\mathbf{C})$ such that

$$F'_\alpha \Phi_{\alpha\beta}(F'_\beta)^{-1} = C_\alpha F_\alpha \Phi_{\alpha\beta} F_\beta^{-1} C_\beta^{-1} \quad \text{in } U_\alpha \cap U_\beta ;$$

recall that flat vector bundles are elements of the cohomology set $H^1(M,GL(m,\mathbf{C}))$. This can evidently be rephrased as the condition that there exist constant matrices C_α such that $T_\alpha = (F'_\alpha)^{-1} C_\alpha F_\alpha$ represents an automorphism of Φ ; or equivalently, that there exists an automorphism $T = (T_\alpha) \in \text{Aut}\,\Phi$ such that $F'_\alpha T_\alpha F_\alpha^{-1}$ are constant matrices. Further, the condition that $F'_\alpha T_\alpha F_\alpha^{-1}$ be constant is just that

$$0 = D(F'_\alpha T_\alpha F_\alpha^{-1})$$

$$= Ad(F_\alpha T_\alpha^{-1})\cdot DF'_\alpha + Ad(F_\alpha)\cdot DT_\alpha + DF_\alpha^{-1}$$

$$= Ad(F_\alpha)\cdot[Ad(T_\alpha^{-1})\cdot\lambda'_\alpha + DT_\alpha - \lambda_\alpha] ,$$

or

$$\lambda'_\alpha = Ad(T_\alpha)\cdot\lambda_\alpha - Ad(T_\alpha)\cdot DT_\alpha = Ad^*(T_\alpha)\cdot\lambda_\alpha ;$$

and by definition, this is equivalent to the condition that λ be equivalent to λ'. Thus, connections yield the same flat representative precisely when they are equivalent, and the theorem is thereby demonstrated.

Remarks. If $\mathfrak{V}$ is a refinement of the covering $\mathfrak{U}$, any refining mapping $\mu: \mathfrak{V} \longrightarrow \mathfrak{U}$ induces a mapping of connections $\mu^*: \Lambda^*(\mathfrak{U},\Phi) \longrightarrow \Lambda^*(\mathfrak{V},\Phi)$; and the correspondence between connections and flat representatives established in Theorem 17 is preserved upon passing to a refinement. Thus it is possible to define connections $\Lambda^*(\Phi)$ by passing to the direct limit over all coverings, and to extend Theorem 17 to a correspondence between $\Lambda^*(\Phi)$ and the set of all flat representatives of Φ. However, by Leray's theorem (see §3(f) of last year's lectures), all flat bundles can be described by $H^1(\mathfrak{U}, GL(m,\mathbb{C}))$ for any covering $\mathfrak{U} = \{U_\alpha\}$ such that all sets U_α and $U_\alpha \cap U_\beta$ are contractible; and thus it is possible to take a fixed covering for all the discussion, to finesse this point altogether.

One slight extension of Theorem 17 should be noted here.

Beginning with a complex analytic vector bundle $\Phi \in \mathcal{V}_m(1) = H^1(M, \mathcal{SL}(m, \mathcal{O}))$, one can consider the set of <u>special flat representatives</u> in the cohomology set $H^1(M, SL(m, \mathbb{C}))$; here $SL(m,*)$ denotes the special linear group, the subgroup $SL(m,*) \subset GL(m,*)$ consisting of matrices of determinant 1, and the natural inclusion

$$i: SL(m, \mathbb{C}) \longrightarrow \mathcal{SL}(m, \mathcal{O})$$

induces the mapping of cohomology sets

$$i^*: H^1(M, SL(m, \mathbb{C})) \longrightarrow H^1(M, \mathcal{SL}(m, \mathcal{O})) .$$

Note that whenever the one cocycle $(\Phi_{\alpha\beta}) \in Z^1(\mathfrak{U}, \mathcal{SL}(m, \mathcal{O}))$ represents the complex analytic vector bundle $\Phi \in \mathcal{V}_m(1)$, it follows that $0 = d \log \det \Phi_{\alpha\beta} = tr(\Phi_{\alpha\beta}^{-1} d\Phi_{\alpha\beta}) = tr(D\Phi_{\alpha\beta})$. If $\lambda = (\lambda_\alpha) \in \Lambda(\mathfrak{U}, \Phi)$ is any complex analytic connection for the bundle Φ, then $\delta(tr\ \lambda) = tr(\delta\lambda) = tr(D\Phi) = 0$, so that $tr\ \lambda \in C^0(\mathfrak{U}, \mathcal{O}(\kappa))$ is actually a section, an element $tr\ \lambda \in \Gamma(M, \mathcal{O}(\kappa))$. A connection $\lambda \in \Lambda(\mathfrak{U}, \Phi)$ will be called a <u>special complex analytic connection</u> if $tr\ \lambda = 0$; and the set of such special connections will be denoted by $\Lambda_0(\mathfrak{U}, \Phi)$. Now if $T = (T_\alpha) \in Aut\ \Phi$, it follows that $\det T_\alpha = \det T_\beta$ in $U_\alpha \cap U_\beta$; so on a compact Riemann surface, $\det T_\alpha$ will be a constant, and so $0 = d \log \det T_\alpha = tr(dT_\alpha T_\alpha^{-1}) = tr(Ad(T_\alpha) \cdot DT_\alpha) = tr\ \tau_\alpha$, where $\tau_\alpha = Ad(T_\alpha) \cdot DT_\alpha$. If $\lambda, \lambda' \in \Lambda(\mathfrak{U}, \Phi)$ are connections, and if $\lambda \sim \lambda'$, then by definition $\lambda' = Ad(T) \cdot \lambda - \tau$; so that $tr\ \lambda' = tr\ Ad(T) \cdot \lambda - tr\ \tau = tr\ \lambda$. The set $\Lambda_0(\mathfrak{U}, \Phi)$ is thus preserved under the equivalence relation introduced on connections; and the equivalence classes of elements of $\Lambda_0(\mathfrak{U}, \Phi)$ will be denoted by $\Lambda_0^*(\mathfrak{U}, \Phi)$.

Corollary. Let M be a compact Riemann surface, and $\Phi \in \mathcal{V}_m(1)$ be a complex analytic vector bundle over M represented by a cocycle $(\Phi_{\alpha\beta}) \in Z^1(\mathcal{U}, \mathcal{S}\mathcal{L}(m, \mathcal{O}))$ in terms of an open covering $\mathcal{U} = \{U_\alpha\}$ of M; and suppose that the covering $\mathcal{U}$ is such that the sets U_α are contractible. Then there is a one-to-one correspondence between the set of special flat representatives of Φ in the covering $\mathcal{U}$, and the set $\Lambda_0^*(\mathcal{U}, \Phi)$ of equivalence classes of special connections for the bundle Φ.

Proof. Following the proof of Theorem 17, if $\lambda = (\lambda_\alpha) \in \Lambda_0(\mathcal{U}, \Phi)$ and if $F_\alpha \in GL(m, \mathcal{O}_{U_\alpha})$ is such that $DF_\alpha = \lambda_\alpha$, then $d \det \log F_\alpha = \operatorname{tr} DF_\alpha = \operatorname{tr} \lambda_\alpha = 0$, so that $\det F_\alpha$ is a constant; since $C_\alpha F_\alpha$ is also a solution of the differential equation $DF_\alpha = \lambda_\alpha$ for any constant matrix $C_\alpha \in GL(m, \mathbf{C})$, it follows that there are solutions $F_\alpha \in SL(m, \mathcal{O}_{U_\alpha})$, and that the most general such solution is of the form $C_\alpha F_\alpha$ for $C_\alpha \in SL(m, \mathbf{C})$. Then $\Phi'_{\alpha\beta} = F_\alpha \Phi_{\alpha\beta} F_\beta^{-1}$ is a special flat bundle, uniquely determined by the special connection λ. Since equivalence in $H^1(M, SL(m, \mathbf{C}))$ and in $H^1(M, GL(m, \mathbf{C}))$ coincides, connections determine the same bundle precisely when they are equivalent, and that serves to complete the proof of the Corollary.

It was noted in Corollary 3 to Theorem 15 that a decomposable complex analytic vector bundle $\Phi = \Phi_1 \oplus \ldots \oplus \Phi_r$ has flat representatives if and only if all of the bundles Φ_i have flat representatives. It should be pointed out that the flat representatives of Φ are not necessarily just the direct sums of the flat representatives of the various bundles Φ_i, though. To illustrate this, consider the particular case of a decomposable vector bundle of rank 2, say $\Phi = \varphi_1 \oplus \varphi_2$, where φ_1, φ_2 are line bundles with $c(\varphi_1) = c(\varphi_2) = 0$, and $\varphi_1 \neq \varphi_2$. Suppose that $(C_{\alpha\beta}) \in Z^1(\mathfrak{U}, GL(2,\mathbf{C}))$ is a flat representative of the bundle Φ, in terms of a coordinate covering $\mathfrak{U} = \{U_\alpha\}$ of the compact Riemann surface M; and that $(C_{\alpha\beta})$ corresponds to the equivalence class of a connection $\lambda = (\lambda_\alpha) \in \Lambda(\mathfrak{U}, \Phi)$ under the correspondence of Theorem 17. Note that the cocycle $(C_{\alpha\beta})$ is decomposable as a direct sum $C_{\alpha\beta} = c_{1\alpha\beta} \oplus c_{2\alpha\beta}$ if and only if the connection $\lambda = (\lambda_\alpha)$ is decomposable correspondingly. (For suppose that the cocycle $(C_{\alpha\beta})$ is decomposable. Recall that under the correspondence of Theorem 17, this cocycle has the form $C_{\alpha\beta} = F_\alpha \Phi_{\alpha\beta} F_\beta^{-1}$, where $(F_\alpha) \in C^0(\mathfrak{U}, \mathcal{GL}(2, \mathcal{O}))$ satisfy $DF_\alpha = \lambda_\alpha$. Explicitly then,

$$\begin{pmatrix} c_{1\alpha\beta} & 0 \\ 0 & c_{2\alpha\beta} \end{pmatrix} \begin{pmatrix} f_{11\beta} & f_{12\beta} \\ f_{21\beta} & f_{22\beta} \end{pmatrix} = \begin{pmatrix} f_{11\alpha} & f_{12\alpha} \\ f_{21\alpha} & f_{22\alpha} \end{pmatrix} \begin{pmatrix} \varphi_{1\alpha\beta} & 0 \\ 0 & \varphi_{2\alpha\beta} \end{pmatrix} ;$$

so that

$$f_{11\alpha} = \frac{c_{1\alpha\beta}}{\varphi_{1\alpha\beta}} f_{11\beta} , \qquad f_{12\alpha} = \frac{c_{1\alpha\beta}}{\varphi_{2\alpha\beta}} f_{12\beta} ,$$

$$f_{21\alpha} = \frac{c_{2\alpha\beta}}{\varphi_{1\alpha\beta}} f_{21\beta} , \qquad f_{22\alpha} = \frac{c_{2\alpha\beta}}{\varphi_{2\alpha\beta}} f_{22\beta} .$$

If the line bundles $(c_{1\alpha\beta})$ and $(\varphi_{1\alpha\beta})$ are equivalent, as can clearly be supposed, it follows immediately that $f_{12\alpha} \equiv 0$ and $f_{21\alpha} \equiv 0$; for $f_{12\alpha}$ and $f_{21\alpha}$ are holomorphic sections of non-trivial complex line bundles of Chern class zero. Then of course

$= \lambda_\alpha$ is decomposable as well. The other direction is trivial.)
s to exhibit the existence of flat representatives of Φ which not decompose into the direct sum of flat representatives of φ_1 φ_2, it suffices to show the existence of complex analytic nections $\lambda \in \Lambda(\mathfrak{U}, \Phi)$ which are not decomposable. These con- tions $\lambda = (\lambda_\alpha)$ are cochains $\lambda \in C^0(\mathfrak{U}, \mathcal{O}(\kappa \otimes \mathrm{Ad}\,\Phi))$ such t $\delta\lambda = D\Phi$, or more explicitly, such that

$$\lambda_\beta - \mathrm{Ad}(\Phi_{\alpha\beta}^{-1})\cdot\lambda_\alpha = D\Phi_{\alpha\beta} ;$$

ting this out in terms of the matrix components $\lambda_{ij\alpha}$, viewed analytic differential forms, the condition is that

$$\begin{pmatrix} {}_{1\beta} & \lambda_{12\beta} \\ {}_{1\beta} & \lambda_{22\beta} \end{pmatrix} - \begin{pmatrix} \varphi_{1\alpha\beta}^{-1} & 0 \\ 0 & \varphi_{2\alpha\beta}^{-1} \end{pmatrix} \begin{pmatrix} \lambda_{11\alpha} & \lambda_{12\alpha} \\ \lambda_{21\alpha} & \lambda_{22\alpha} \end{pmatrix} \begin{pmatrix} \varphi_{1\alpha\beta} & 0 \\ 0 & \varphi_{2\alpha\beta} \end{pmatrix} = \begin{pmatrix} D\varphi_{1\alpha\beta} & 0 \\ 0 & D\varphi_{2\alpha\beta} \end{pmatrix},$$

ce that

$$\lambda_{11\alpha} = \lambda_{11\beta} - D\varphi_{1\alpha\beta} , \qquad \lambda_{12\alpha} = \frac{\varphi_{1\alpha\beta}}{\varphi_{2\alpha\beta}} \lambda_{12\beta} ,$$

$$\lambda_{21\alpha} = \frac{\varphi_{2\alpha\beta}}{\varphi_{1\alpha\beta}} \lambda_{21\beta} , \qquad \lambda_{22\alpha} = \lambda_{22\beta} - D\varphi_{2\alpha\beta} .$$

s λ_{ii} is any complex analytic connection for the complex line dle φ_i ; while $\lambda_{12\alpha}$ and $\lambda_{21\alpha}$ are sections,

$$(\lambda_{12\alpha}) \in \Gamma(M, \mathcal{O}^{1,0}(\varphi_1\varphi_2^{-1})) ,$$

$$(\lambda_{21\alpha}) \in \Gamma(M, \mathcal{O}^{1,0}(\varphi_2\varphi_1^{-1})) .$$

ce $\varphi_1 \neq \varphi_2$ but $c(\varphi_1) = c(\varphi_2) = 0$, it follows from the Riemann-

Roch theorem that $\dim \Gamma(M, \mathcal{O}^{1,0}(\varphi_1\varphi_2^{-1})) = \dim \Gamma(M, \mathcal{O}^{1,0}(\varphi_2\varphi_1^{-1})) = g-1$; and therefore, on a Riemann surface of genus $g > 1$, there are connections $\lambda \in \Lambda(\mathfrak{U}, \varphi_1 \oplus \varphi_2)$ which are not decomposable. (Since the fundamental group of a surface of genus 1 is abelian, all flat bundles are necessarily decomposable in that case; recall the discussion in §9(b) of last year's lectures. The restriction $g > 1$ in the preceding observation is thus to be expected.)

§7. Flat sheaves: geometric aspects.

(a) The last chapter was devoted to a discussion of the complex analytic vector bundles which admit flat representatives; the present chapter will take up the discussion of flat vector bundles themselves.

Paralleling the terminology introduced earlier for complex analytic vector bundles (see §4(a)), a flat vector bundle $\chi_1 \in H^1(M, GL(n_1, \mathbf{C}))$ is called a subbundle of a flat vector bundle $\chi \in H^1(M, GL(n, \mathbf{C}))$ if there are representative cocycles $(\chi_{1\alpha\beta}) \in Z^1(\mathfrak{U}, GL(n_1, \mathbf{C}))$ and $(\chi_{\alpha\beta}) \in Z^1(\mathfrak{U}, GL(n, \mathbf{C}))$ of the form

$$\chi_{\alpha\beta} = \begin{pmatrix} \chi_{1\alpha\beta} & \chi_{12\alpha\beta} \\ 0 & \chi_{2\alpha\beta} \end{pmatrix} \tag{1}$$

for some covering $\mathfrak{U}$ of the space M ; the cocycle $(\chi_{2\alpha\beta}) \in Z^1(\mathfrak{U}, GL(n_2, \mathbf{C}))$ represents the quotient bundle $\chi_2 = \chi/\chi_1 \in H^1(M, GL(n_2, \mathbf{C}))$. A flat vector bundle χ is called reducible if it contains a proper flat subbundle; otherwise χ is called irreducible. A flat vector bundle χ is said to be decomposable into the direct sum of two flat vector bundles χ_1 and χ_2, written $\chi = \chi_1 \oplus \chi_2$, if there are representative cocycles of the form (1) in which $\chi_{12\alpha\beta} = 0$; if no such decomposition is possible, the bundle is said to be indecomposable. Again, a decomposable bundle is reducible, but the converse does not generally hold.

Recall from last year's lectures (pages 184-189) that to each flat vector bundle $\chi \in H^1(M, GL(n, \mathbf{C}))$ there is associated a homomorphism $\hat{\chi}: \pi_1(M) \longrightarrow GL(n, \mathbf{C})$ called the characteristic representation of that vector bundle, and that $\hat{\chi}$ is uniquely determined up to an inner automorphism in $GL(n, \mathbf{C})$; here $\pi_1(M)$ denotes the fundamental group of the surface, with a fixed but arbitrary base point. The mapping $X \longrightarrow \hat{X}$ establishes a one-to-one correspondence

$$(2) \qquad H^1(M, GL(n,\mathbb{C})) \longrightarrow \mathrm{Hom}(\pi_1(M), GL(n,\mathbb{C}))/GL(n,\mathbb{C}) ,$$

where the quotient on the right denotes equivalence classes of homomorphisms under inner automorphisms of $GL(n,\mathbb{C})$. In particular, every flat vector bundle over a simply-connected surface M is trivial; or more generally, the restriction of a flat vector bundle to any simply-connected open subset of M is a trivial vector bundle over that subset. It is also easy to see that a flat vector bundle $X \in H^1(M, GL(n,\mathbb{C}))$ is irreducible (or indecomposable) precisely when its characteristic representation $\hat{X} \in \mathrm{Hom}(\pi_1(M), GL(n,\mathbb{C}))$ is irreducible (or indecomposable, respectively); details will be left to the readers.

A flat vector bundle X represents a complex analytic vector bundle, so the sheaves $\mathcal{E}(X)$ of germs of C^∞ sections of X and $\mathcal{O}(X)$ of germs of holomorphic sections of X are defined as usual. Moreover, since the bundle X has a representative cocycle consisting entirely of constant matrices, the subset of $\mathcal{O}(X)$ consisting of germs of locally constant functions is a non-trivial subsheaf of $\mathcal{O}(X)$; this will be called the <u>sheaf of germs of flat sections</u> of the bundle X, and will be denoted by $\mathcal{F}(X)$. (Note that the sheaf $\mathcal{F}(X)$ can thus be defined as follows. Select a basis $\mathfrak{U} = \{U_\alpha\}$ of the open sets of M, and a representative cocycle $(X_{\alpha\beta}) \in Z^1(\mathfrak{U}, GL(n,\mathbb{C}))$ for the flat vector bundle X. To each open set $U_\alpha \in \mathfrak{U}$ associate the complex vector space $\mathcal{F}_\alpha = \mathbb{C}^n$, viewed as the space of column vectors of length n;

and to each inclusion relation $U_\alpha \subset U_\beta$ associate the linear transformation $\chi_{\alpha\beta}: \mathcal{F}_\beta \longrightarrow \mathcal{F}_\alpha$ defined by the matrix $\chi_{\alpha\beta}$. It is readily seen that the collection $\{U_\alpha, \mathcal{F}_\alpha, \chi_{\alpha\beta}\}$ is a complete presheaf of complex vector spaces over M; the associated sheaf is defined to be the sheaf $\mathcal{F}(\chi)$.) In general, a sheaf $\mathcal{F}$ of complex vector spaces over M will be called a <u>flat sheaf of rank</u> n if each point $p \in M$ has an open neighborhood U such that the restriction of $\mathcal{F}$ to U is the trivial sheaf of vector spaces of rank n over U, that is, such that $\mathcal{F}|U \cong \mathbb{C}^n \times U$. The sheaf of germs of flat sections of a flat vector bundle is obviously a flat sheaf; and conversely, it is clear that any flat sheaf $\mathcal{F}$ is the sheaf of germs of flat sections of a flat vector bundle which will be denoted by $\chi(\mathcal{F})$.

Of course, this is merely another special case of the general structure of sheaves of modules over sheaves of rings, as discussed in §1. Taking for $\mathcal{R}$ the trivial sheaf of rings* $\mathcal{R} = \mathbf{C} \times M$ over M, the flat sheaves of rank n are just the locally free sheaves of $\mathcal{R}$-modules of rank n; and as in Theorem 1, such sheaves are described by their representative fibre bundles in $H^1(M, \mathcal{GL}(n, \mathcal{R})) = H^1(M, GL(n, \mathbf{C}))$. The notation is a bit different from that introduced in the earlier general discussion, as a matter of convenience only. Flat sheaves can also

*Note that $\mathbf{C} \times M$ viewed as a sheaf has the product topology of the discrete topology on $\mathbf{C}$ and the usual topology on M.

be viewed as sheaves of local coefficients in the sense of Steenrod; (see N. E. Steenrod, The Topology of Fibre Bundles, (Princeton University Press, 1951), especially §30).

(b) For a simply-connected manifold M any flat vector bundle X is trivial, hence $H^q(M, \mathcal{F}(X)) \cong H^q(M, \mathbb{C}^n)$ for all dimensions q ; thus for a contractible manifold M it follows that $H^q(M, \mathcal{F}(X)) = 0$ for all dimensions $q > 0$. A compact manifold M admits a finite open covering $\mathcal{U} = \{U_\alpha\}$ such that the sets U_α and all their intersections are contractible; and by the above remark, such a covering is a Leray covering of M for the sheaf $\mathcal{F}(X)$, so that $H^q(\mathcal{U}, \mathcal{F}(X)) \cong H^q(M, \mathcal{F}(X))$ for all dimensions q. Now a q-cochain in $C^q(\mathcal{U}, \mathcal{F}(X))$ is given by an n-dimensional complex vector associated to each intersection $U_{\alpha_0} \cap \ldots \cap U_{\alpha_q} \neq 0$; so it is evident that $C^q(\mathcal{U}, \mathcal{F}(X))$ is a finite-dimensional complex vector space. Thus, <u>for a compact manifold M, all the cohomology groups $H^q(M, \mathcal{F}(X))$ are finite-dimensional complex vector spaces</u>, whenever $\mathcal{F}(X)$ is a sheaf of germs of flat sections of a flat vector bundle X. The lowest cohomology groups can be described explicitly quite easily.

<u>Theorem 18.</u> Let $X \in H^1(M, GL(n,\mathbb{C}))$ be a flat vector bundle over any (connected) manifold M. If X is irreducible, then $H^0(M, \mathcal{F}(X)) = 0$; and more generally, $H^0(M, \mathcal{F}(X)) \cong \mathbb{C}^r$ where r is the largest integer such that the trivial bundle of rank r is contained as a subbundle of X.

Proof. Let $\mathcal{U} = \{U_\alpha\}$ be an open covering of M such that the bundle X is defined by a cocycle $(X_{\alpha\beta}) \in Z^1(\mathcal{U}, GL(n,\mathbb{C}))$.

If $H^0(M, \mathcal{F}(\chi)) = \Gamma(M, \mathcal{F}(\chi)) = \mathbb{C}^r$, select r linearly independent sections $F_i \in \Gamma(M, \mathcal{F}(\chi))$; each section F_i is given by a column vector $F_{i\alpha}$ associated to each set U_α such that $F_{i\alpha} = \chi_{\alpha\beta}\cdot F_{i\beta}$ whenever $U_\alpha \cap U_\beta \neq \emptyset$. For each U_α the r vectors $F_{i\alpha}$, $i = 1,\ldots,r$, are linearly independent; so there is a nonsingular matrix $\Theta_\alpha \in GL(n,\mathbb{C})$ such that $\Theta_\alpha\cdot E_i = F_{i\alpha}$, where E_i is the column vector having the entry 1 in the i-th place and entries 0 elsewhere. The cocycle $\chi'_{\alpha\beta} = \Theta_\alpha^{-1}\, \chi_{\alpha\beta}\, \Theta_\beta$ is equivalent to the cocycle $\chi_{\alpha\beta}$, hence defines the same flat bundle χ . Since $\chi'_{\alpha\beta}\cdot E_i = E_i$ for $i = 1,\ldots,r$, however, it follows that the matrices $\chi'_{\alpha\beta}$ have the form

$$\chi'_{\alpha\beta} = \begin{pmatrix} 1 & 0 & \ldots & 0 & * & \ldots & * \\ 0 & 1 & \ldots & 0 & * & \ldots & * \\ & \ldots & & & \ldots & & \ldots \\ 0 & 0 & \ldots & 1 & * & \ldots & * \\ 0 & 0 & \ldots & 0 & * & \ldots & * \\ & \ldots & & & \ldots & & \ldots \\ 0 & 0 & \ldots & 0 & * & \ldots & * \end{pmatrix} \begin{matrix} \left.\vphantom{\begin{matrix}1\\1\\1\\1\end{matrix}}\right\} i \\ \left.\vphantom{\begin{matrix}1\\1\\1\end{matrix}}\right\} n-i \end{matrix} \quad ;$$

thus the identity bundle of rank r , the bundle defined by the cocycle formed of identity matrices of rank r , appears as a subbundle of χ . In particular, if χ is irreducible, necessarily $r = 0$. That concludes the proof.

The corresponding one-dimensional cohomology group can be described purely algebraically in terms of the characteristic representation; the machinery involved is the cohomology theory of abstract groups, but a very simple-minded approach is quite adequate. Selecting a particular mapping $\hat{\chi}: \pi_1(M) \longrightarrow GL(n,\mathbb{C})$ for the characteristic representation, $\hat{X}$ can be viewed as exhibiting $\pi_1(M)$ as a group of operators on the vector space $\mathbb{C}^n$, or what is the same thing, $\mathbb{C}^n$ is a $\pi_1(M)$-module under the action described by $\hat{X}$. A 1-cocycle of the group $\pi_1(M)$ with coefficients in this $\pi_1(M)$-module is a mapping $A: \pi_1(M) \longrightarrow \mathbb{C}^n$, associating to each element $\gamma \in \pi_1(M)$ a column vector $A_\gamma \in \mathbb{C}^n$, such that

$$(3) \qquad A_{\gamma_1\gamma_2} = \hat{X}_{\gamma_2}^{-1} \cdot A_{\gamma_1} + A_{\gamma_2} \quad \text{for all} \quad \gamma_1, \gamma_2 \in \pi_1(M) \ ;$$

the set of these cocycles form a complex vector space denoted by $Z^1(\pi_1(M),\hat{X})$. A 1-cocycle $A \in Z^1(\pi_1(M),\hat{X})$ is called a 1-coboundary if there is a vector $B \in \mathbb{C}^n$ such that

$$(4) \qquad A_\gamma = B - \hat{X}_\gamma^{-1} \cdot B \quad \text{for all} \quad \gamma \in \pi_1(M) \ ;$$

it is easy to see that (4) defines a 1-cocycle A for any vector $B \in \mathbb{C}^n$, and that thus the set of 1-coboundaries forms a vector subspace $B^1(\pi_1(M),\hat{X})$. The quotient space

$$H^1(\pi_1(M),\hat{X}) = Z^1(\pi_1(M),\hat{X})/B^1(\pi_1(M),\hat{X})$$

is called the first cohomology group of the group $\pi_1(M)$ with coefficients in the $\pi_1(M)$-module associated to the representation $\hat{X}$. It is easy to see further that this cohomology group is unchanged when the mapping $\hat{X}$ is replaced by a conjugate representation, in the sense that the two cohomology groups are canonically isomorphic; thus the cohomology group can be considered as associated to the characteristic representation itself.

<u>Theorem 19.</u> Let $X \in H^1(M,GL(n,\mathbb{C}))$ be a flat vector bundle over any (connected) manifold M, with characteristic representation

$\hat{X}$. Then there is a natural isomorphism

$$H^1(M, \mathcal{F}(X)) \cong H^1(\pi_1(M), \hat{X}) .$$

Proof. Let $\mathfrak{U} = \{U_\alpha\}$ be a Leray covering of the manifold M for the sheaf $\mathcal{F}(X)$; it is apparent that the covering can also be so selected that $\pi_1(M) = \pi_1(\mathfrak{U}, U_o)$, where $U_o \in \mathfrak{U}$ is a base for the chains of the covering $\mathfrak{U}$, and that the bundle X has a representative cocycle $(X_{\alpha\beta}) \in Z^1(\mathfrak{U}, GL(n, \mathbb{C}))$. (For the terminology and notation, see pages 184-189 of last year's lectures.) Consider an arbitrary cocycle $A = (A_{\alpha\beta}) \in Z^1(\mathfrak{U}, \mathcal{F}(X))$; thus $A_{\alpha\beta} \in \mathbb{C}^n$ are vectors (viewed as column vectors) associated to intersections $U_\alpha \cap U_\beta \neq \emptyset$, such that $A_{\alpha\gamma} = X^{-1}_{\beta\gamma} \cdot A_{\alpha\beta} + A_{\beta\gamma}$ whenever $U_\alpha \cap U_\beta \cap U_\gamma \neq \emptyset$. To each closed chain $\gamma = (U_{\alpha_o}, U_{\alpha_1}, \ldots, U_{\alpha_p})$ of the covering $\mathfrak{U}$ based at U_o associate the vector

$$\hat{A}_\gamma = (X_{\alpha_1\alpha_2} \cdots X_{\alpha_{p-1}\alpha_p})^{-1} A_{\alpha_o\alpha_1} + (X_{\alpha_2\alpha_3} \cdots X_{\alpha_{p-1}\alpha_p})^{-1} A_{\alpha_1\alpha_2} + \cdots$$
$$\cdots + X^{-1}_{\alpha_{p-1}\alpha_p} A_{\alpha_{p-2}\alpha_{p-1}} + A_{\alpha_{p-1}\alpha_p} .$$

Note that $\hat{A}_{\gamma_1} = \hat{A}_\gamma$ whenever γ_1 and γ are homotopic chains. (It suffices to show this when γ_1 results from a simple jerk on the chain γ. Suppose then that $U_{\alpha_{i-1}} \cap U_{\alpha_i} \cap U_\beta \neq \emptyset$, and that γ_1 is the closed chain $\gamma_1 = (U_{\alpha_o}, \ldots, U_{\alpha_{i-1}}, U_\beta, U_{\alpha_i}, \ldots, U_{\alpha_p})$. The fact that $\hat{A}_{\gamma_1} = \hat{A}_\gamma$ in this case follows immediately from the cocycle condition; details will be left to the readers.) Thus $\hat{A}$ can be viewed as a mapping $\hat{A}: \pi_1(\mathfrak{U}, U_o) \longrightarrow \mathbb{C}^n$. If γ_1, γ_2 are two closed chains, it follows by an easy calculation that

$$\hat{A}_{\gamma_1\gamma_2} = \hat{X}_{\gamma_2}^{-1} \cdot \hat{A}_{\gamma_1} + \hat{A}_{\gamma_2} ;$$

thus actually $\hat{A} \in Z^1(\pi_1(\mathfrak{U}, U_o), \hat{X})$. The mapping $A \longrightarrow \hat{A}$ is a well-defined linear mapping from the vector space $Z^1(\mathfrak{U}, \mathcal{F}(X))$ to the vector space $Z^1(\pi_1(\mathfrak{U}, U_o), \hat{X})$; we shall see that it establishes the desired isomorphism. First suppose that the cocyle A is cohomologous to zero, that is, that there is a zero-cochain $B = (B_\alpha) \in C^o(\mathfrak{U}, \mathcal{F}(X))$ such that $A_{\alpha\beta} = B_\beta - X_{\alpha\beta}^{-1} \cdot B_\alpha$ whenever $U_\alpha \cap U_\beta \neq \emptyset$. It follows readily that then for any closed chain $\gamma = (U_{\alpha_o}, \ldots, U_{\alpha_p})$ the mapping $\hat{A}$ has the value $\hat{A}_\gamma = B_{\alpha_o} - \hat{X}_\gamma^{-1} \cdot B_{\alpha_o}$, so that $\hat{A}$ is also a coboundary. The mapping $A \longrightarrow \hat{A}$ thus induces a well defined linear mapping from $H^1(\mathfrak{U}, \mathcal{F}(X))$ into $H^1(\pi_1(\mathfrak{U}, U_o), \hat{X})$. Next suppose that for some cocycle $A \in Z^1(\mathfrak{U}, \mathcal{F}(X))$ the image cocycle $\hat{A} \in Z^1(\pi_1(\mathfrak{U}, U_o), \hat{X})$ is cohomologous to zero, that is, that there is a vector $B \in \mathbb{C}^n$ such that $\hat{A}_\gamma = B - \hat{X}_\gamma^{-1} \cdot B$ for every closed chain γ of the covering $\mathfrak{U}$. For any element $U_\alpha \in \mathfrak{U}$ select a chain $\pi(\alpha)$ based at U_o and ending at U_α; and if $\pi(\alpha) = (U_{\alpha_o}, U_{\alpha_1}, \ldots, U_{\alpha_p})$ with $U_{\alpha_p} = U_\alpha$, set

$$B_\alpha = (X_{\alpha_o\alpha_1} \cdots X_{\alpha_{p-1}\alpha_p})^{-1} B + (X_{\alpha_1\alpha_2} \cdots X_{\alpha_{p-1}\alpha_p})^{-1} A_{\alpha_o\alpha_1} + \cdots$$

$$\cdots + X_{\alpha_{p-1}\alpha_p}^{-1} A_{\alpha_{p-2}\alpha_{p-1}} + A_{\alpha_{p-1}\alpha_p} .$$

It is a straightforward matter to verify that B_α so defined is independent of the choice of the chain $\pi(\alpha)$; and that the cochain $B = (B_\alpha) \in C^o(\mathfrak{U}, \mathcal{F}(X))$ has the cocycle A as its coboundary, so that A is cohomologous to zero. Thus the mapping from

$H^1(\mathfrak{U}, \mathcal{F}(X))$ to $H^1(\pi_1(\mathfrak{U}, U_o), \hat{X})$ is an isomorphism into. Finally, suppose given an arbitrary one-cocycle $\hat{A} = (\hat{A}_\alpha) \in Z^1(\pi_1(\mathfrak{U}, U_o), \hat{X})$. For any element $U_\alpha \in \mathfrak{U}$ select a chain $\pi(\alpha)$ based at U_o and ending at U_α; and set $X_{\pi(\alpha)} = X_{\alpha_0\alpha_1} \cdot X_{\alpha_1\alpha_2} \cdot \ldots \cdot X_{\alpha_{p-1}\alpha_p}$ where $\pi(\alpha) = (U_{\alpha_0}, U_{\alpha_1}, \ldots, U_{\alpha_p})$. If $U_\alpha \cap U_\beta \neq \emptyset$ for two subsets $U_\alpha, U_\beta \in \mathfrak{U}$, then the chain $\pi(\alpha)\cdot\pi(\beta)^{-1}$ is a closed chain based at U_o; set

$$A_{\alpha\beta} = X^{-1}_{\pi(\beta)} \cdot \hat{A}_{\pi(\alpha)\cdot\pi(\beta)^{-1}} .$$

Since $\hat{A}$ is a cocycle, it follows readily that whenever $U_\alpha \cap U_\beta \cap U_\gamma \neq \emptyset$ the element

$$A_{\alpha\gamma} = X^{-1}_{\pi(\gamma)} \cdot \hat{A}_{\pi(\alpha)\cdot\pi(\gamma)^{-1}} = X^{-1}_{\pi(\gamma)} \hat{A}_{\pi(\alpha)\cdot\pi(\beta)^{-1}\cdot\pi(\beta)\cdot\pi(\gamma)^{-1}} =$$

$$= X^{-1}_{\pi(\gamma)} \left[\hat{X}^{-1}_{\pi(\beta)\cdot\pi(\gamma)^{-1}} \cdot \hat{A}_{\pi(\alpha)\cdot\pi(\beta)^{-1}} + \hat{A}_{\pi(\beta)\cdot\pi(\gamma)^{-1}} \right] = X^{-1}_{\beta\gamma} \cdot A_{\alpha\beta} + A_{\beta\gamma} ;$$

so the vectors $A_{\alpha\beta}$ form a one-cocycle $A = (A_{\alpha\beta}) \in Z^1(\mathfrak{U}, \mathcal{F}(X))$. It is easily verified that this cocycle maps onto the given cohomology class $\hat{A}$ under the isomorphism from $H^1(\mathfrak{U}, \mathcal{F}(X))$ to $H^1(\pi_1(\mathfrak{U}, U_o), \hat{X})$; the latter isomorphism is therefore onto, and the proof is thereby concluded.

An application of Theorem 19 leads to a simple explicit calculation of the cohomology group $H^1(M, \mathcal{F}(X))$ over a compact Riemann surface M. Recall that the fundamental group $\pi_1(M)$ of a compact orientable surface M of genus g can be described as a group with $2g$ generators $\sigma_1, \ldots, \sigma_g, \tau_1, \ldots, \tau_g$ and the one relation

$$(5) \qquad \sigma_g\tau_g\sigma_g^{-1}\tau_g^{-1} \cdots \sigma_1\tau_1\sigma_1^{-1}\tau_1^{-1} = 1 \; ;$$

(see for instance H. Seifert and W. Threlfall, Lehrbuch der Topologie (Chelsea, 1947)). Selecting any homomorphism $\hat{\chi}: \pi_1(M) \longrightarrow GL(n,\mathbb{C})$ as the characteristic representation of a flat vector bundle $\chi \in H^1(M,GL(n,\mathbb{C}))$ over the surface M, a one-cocycle $A \in Z^1(\pi_1(M),\hat{\chi})$ is described by any $2g$ vectors $A_{\sigma_i}, A_{\tau_i} \in \mathbb{C}^n$ subject to one relation corresponding to the group relation (5). To make this explicit, note from (3) that $A_{\gamma^{-1}} = -\hat{\chi}_\gamma \cdot A_\gamma$ for any $\gamma \in \pi_1(M)$; and setting

$$[\sigma,\tau] = \sigma\tau\sigma^{-1}\tau^{-1} \quad \text{for any} \quad \sigma,\tau \in \pi_1(M) \; ,$$

it further follows from (3) that

$$(6) \qquad \begin{aligned} A_{[\sigma,\tau]} &= \hat{\chi}_{\tau\sigma\tau^{-1}} \cdot A_\sigma + \hat{\chi}_{\tau\sigma} \cdot A_\tau - \hat{\chi}_{\tau\sigma} \cdot A_\sigma - \hat{\chi}_\tau \cdot A_\tau \\ &= \hat{\chi}_{\tau\sigma} \cdot [(I - \hat{\chi}_\sigma^{-1}) \cdot A_\tau - (I - \hat{\chi}_\tau^{-1}) \cdot A_\sigma] \; , \end{aligned}$$

where I denotes the identity matrix. Thus the condition that the vectors $A_{\sigma_i}, A_{\tau_i} \in \mathbb{C}^n$ form a one-cocycle is just that

$$(7) \qquad \begin{aligned} 0 &= A_{[\sigma_g,\tau_g] \cdots [\sigma_1,\tau_1]} \\ &= \hat{\chi}^{-1}_{[\sigma_{g-1},\tau_{g-1}]\cdots[\sigma_1,\tau_1]} \cdot A_{[\sigma_g,\tau_g]} + \cdots + \hat{\chi}^{-1}_{[\sigma_1,\tau_1]} \cdot A_{[\sigma_2,\tau_2]} + A_{[\sigma_1,\tau_1]} \end{aligned}$$

where the terms $A_{[\sigma_i,\tau_i]}$ are given by equation (6). Such a one-cocycle is a one-coboundary if and only if there is a vector $B \in \mathbb{C}^n$ such that

$$(8) \qquad A_{\sigma_i} = B - \hat{\chi}^{-1}_{\sigma_i} \cdot B, \qquad A_{\tau_i} = B - \hat{\chi}^{-1}_{\tau_i} \cdot B$$

for $i = 1,\ldots,g$.

<u>Corollary.</u> Let $X \in H^1(M, GL(n,\mathbb{C}))$ be a flat vector bundle over a compact Riemann surface of genus g. Then $\dim_{\mathbb{C}} H^1(M, \mathcal{F}(X)) = 2(g-1)n+p+q$, where p is the largest integer such that the trivial bundle of rank p is contained as a subbundle of X, and q is the largest integer such that the trivial bundle of rank q is contained as a subbundle of the dual vector bundle X^*.

Proof. The condition that the vectors A_{σ_i}, A_{τ_i} determine a one-cocycle $A \in Z^1(\pi_1(M), \hat{X})$, as expressed in equation (7), is a system of n linear constraints on the $2gn$ components of those vectors; and so $\dim_{\mathbb{C}} Z^1(\pi_1(M), \hat{X}) = 2gn - r$, where r is the rank of the system of linear equations (7). Viewing that system of equations as being of the form

$$\sum_{i=1}^{g} (\Theta_{\sigma_i} \cdot A_{\sigma_i} + \Theta_{\tau_i} \cdot A_{\tau_i}) = 0$$

for suitable matrices $\Theta_{\sigma_i}, \Theta_{\tau_i}$, the corank $n-r$ is the dimension of the space of row vectors $R \in \mathbb{C}^n$ such that $R \cdot \Theta_{\sigma_i} = R \cdot \Theta_{\tau_i} = 0$ for $i = 1, \ldots, g$. For any row vector $R \in \mathbb{C}^n$ set $R_i = R \cdot \hat{X}^{-1}_{[\sigma_{i-1}, \tau_{i-1}] \ldots [\sigma_1, \tau_1]}$; thus $R_1 = R$, and $R_i \cdot \hat{X}^{-1}_{[\sigma_i, \tau_i]} = R_i \cdot \hat{X}_{\tau_i} \hat{X}_{\sigma_i} \hat{X}^{-1}_{\tau_i} \hat{X}^{-1}_{\sigma_i} = R_{i+1}$. Now the conditions $R \cdot \Theta_{\sigma_i} = R \cdot \Theta_{\tau_i} = 0$ are readily seen to be of the form

$$R_i \cdot \hat{X}_{\tau_i \sigma_i} (I - \hat{X}^{-1}_{\sigma_i}) = R_i \cdot \hat{X}_{\tau_i \sigma_i} (I - \hat{X}^{-1}_{\tau_i}) = 0 ,$$

or equivalently

$$R_i \cdot \hat{X}_{\tau_i} = R_i \cdot \hat{X}_{\tau_i} \hat{X}_{\sigma_i} = R_i \cdot \hat{X}_{\tau_i} \hat{X}_{\sigma_i} \hat{X}^{-1}_{\tau_i} ;$$

but this clearly implies that $R_i = R_i \cdot \hat{X}_{\tau_i} = R_i \cdot \hat{X}_{\sigma_i}$, and hence also that $R_i = R_{i+1}$. Therefore whenever the row vector R satisfies $R \cdot \Theta_{\sigma_i} = R \cdot \Theta_{\tau_i} = 0$, the column vector ${}^tR \in \mathbb{C}^n$ is such that

$$ {}^t\hat{X}_\gamma^{-1} \cdot {}^tR = {}^tR \qquad \text{for all} \quad \gamma \in \pi_1(M) . $$

Consequently the corank $q = n-r$ is the largest integer q such that the identity representation of rank q is contained in ${}^t\hat{X}^{-1}$, or equivalently, the largest integer q such that the trivial bundle of rank q is contained as a subbundle of the dual vector bundle X^* .

Now the coboundaries $B^1(\pi_1(M),\hat{X})$ are just the cocycles of the form $A_\gamma = B - \hat{X}_\gamma^{-1} \cdot B$ for any vector $B \in \mathbb{C}^n$. Thus $\dim_{\mathbb{C}} B^1(\pi_1(M),\hat{X}) = n-p$, where p is the dimension of the subspace of all those vectors $B \in \mathbb{C}^n$ such that $\hat{X}_\gamma \cdot B = B$ for all $\gamma \in \pi_1(M)$; equivalently of course, p is the largest integer such that the identity representation of rank p is contained in $\hat{X}$, or the largest integer such that the trivial bundle of rank p is contained as a subbundle of the vector bundle X . It therefore follows that $\dim H^1(M, \mathcal{F}(X)) = \dim H^1(\pi_1(M),\hat{X}) = \dim Z^1(\pi_1(M),\hat{X}) - \dim B^1(\pi_1(M),\hat{X}) = 2(g-1)n+p+q$, which is the desired result.

Remarks. Note that saying that the trivial bundle of rank q is contained as a subbundle of the dual vector bundle X^* is equivalent to saying that the trivial vector bundle of rank q is a quotient bundle of the vector bundle X itself. Note also that if X is the trivial bundle of rank n , then $\mathcal{F}(X) \cong \mathbb{C}^n \times M$; Theorem 18 and the Corollary to Theorem 19 assert that $H^0(M, \mathcal{F}(X)) \cong \mathbb{C}^n$ and $H^1(M, \mathcal{F}(X)) \cong \mathbb{C}^{2gn}$, as they should.

(c) There is a form of the deRham isomorphism which can be used to describe the cohomology groups of a manifold with coefficients in the sheaf of germs of flat sections of a flat vector bundle. (For the usual deRham isomorphism, and the required background material about differential forms, see §5(a) of last year's lectures.) If $\mathcal{E}^p(X)$ denotes the sheaf of germs of C^∞ differential forms which are sections of the flat vector bundle $X \in H^1(M, GL(n, \mathbb{C}))$ over the Riemann surface M, the operation of exterior differentiation is a well-defined sheaf homomorphism $d: \mathcal{E}^p(X) \longrightarrow \mathcal{E}^{p+1}(X)$. There then arises the exact sequence of sheaves (the <u>deRham sequence</u>)

$$(9) \quad 0 \longrightarrow \mathcal{F}(X) \longrightarrow \mathcal{E}^0(X) \xrightarrow{d} \mathcal{E}^1(X) \xrightarrow{d} \mathcal{E}^2(X) \longrightarrow 0 ,$$

recalling that M is two-dimensional. The sheaves $\mathcal{E}^p(X)$ are locally isomorphic to $\mathcal{E}^p$, hence are fine sheaves; so letting $\mathcal{E}^1_c(X) \subset \mathcal{E}^1(X)$ be the subsheaf which is the kernel of d, (the subsheaf of closed differential forms), it follows that

$$(10) \quad \begin{cases} H^1(M, \mathcal{F}(X)) \cong \Gamma(M, \mathcal{E}^1_c(X))/d\Gamma(M, \mathcal{E}^0(X)) , \\ H^2(M, \mathcal{F}(X)) \cong \Gamma(M, \mathcal{E}^2(X))/d\Gamma(M, \mathcal{E}^1(X)) , \\ H^q(M, \mathcal{F}(X)) = 0 \quad \text{whenever } q \geq 3 . \end{cases}$$

(See Theorem 3 of last year's lectures for the proof of these assertions; the isomorphisms (10) will be called the deRham isomorphisms, as a convenient abbreviation.) Given any differential form in $\Gamma(M, \mathcal{E}^r_c(X))$, the cohomology class associated to that differential form by means of the isomorphism (10) is called the <u>period class</u> of the differential form.

This deRham isomorphism is particularly useful in describing a duality for cohomology with coefficients in the sheaves $\mathcal{F}(\chi)$. If $\chi \in H^1(M, GL(n,\mathbb{C}))$ is a flat vector bundle over the Riemann surface M, let $\overline{\chi}^*$ be the complex conjugate of its dual bundle; so if χ is defined by a one-cocycle $(\chi_{\alpha\beta}) \in Z^1(\mathfrak{U}, GL(n,\mathbb{C}))$, then $\overline{\chi}^*$ is the flat vector bundle defined by the one-cocycle $({}^t\overline{\chi}_{\alpha\beta}^{-1}) \in Z^1(\mathfrak{U}, GL(n,\mathbb{C}))$. By the deRham isomorphism (10), any cohomology class $A \in H^p(M, \mathcal{F}(\chi))$ can be represented by a closed differential form $\varphi \in \Gamma(M, \mathcal{E}^p(\chi))$; this representing differential form is not unique, but the most general such is given by $\varphi + d\varphi'$ for arbitrary differential forms $\varphi' \in \Gamma(M, \mathcal{E}^{p-1}(\chi))$. Recall that actually φ is given by vectors of differential forms φ_α in the various open sets U_α of a suitable open covering of the surface, such that $\varphi_\alpha = \chi_{\alpha\beta} \cdot \varphi_\beta$ in each intersection $U_\alpha \cap U_\beta$; the vectors φ_α are viewed as column vectors, as usual. In a similar manner, any cohomology class $B \in H^{2-p}(M, \mathcal{F}(\overline{\chi}^*))$ can be represented by a closed differential form $\psi \in \Gamma(M, \mathcal{E}^{2-p}(\overline{\chi}^*))$, or more generally by the closed differential forms $\psi + d\psi'$ for any $\psi' \in \Gamma(M, \mathcal{E}^{1-p}(\overline{\chi}^*))$. The exterior product ${}^t\overline{\psi} \wedge \varphi$ is then a global scalar differential form of degree 2 on the Riemann surface M; for in $U_\alpha \cap U_\beta$ it is evident that ${}^t\overline{\psi}_\alpha \wedge \varphi_\alpha = {}^t\overline{\psi}_\beta \cdot \chi_{\alpha\beta}^{-1} \wedge \chi_{\alpha\beta}\varphi_\beta = {}^t\overline{\psi}_\beta \wedge \varphi_\beta$. Upon choosing different representative differential forms for the same cohomology classes, the exterior product form is modified to become ${}^t(\overline{\psi} + d\overline{\psi}') \wedge (\varphi + d\varphi') = {}^t\overline{\psi} \wedge \varphi + d({}^t\overline{\psi} \wedge \varphi' + {}^t\overline{\psi}' \wedge \varphi + {}^t\overline{\psi}' \wedge d\varphi')$; thus the class of the form ${}^t\overline{\psi} \wedge \varphi$ in the quotient space $\Gamma(M, \mathcal{E}^2)/d\Gamma(M, \mathcal{E}^1)$ depends only on the original cohomology classes.

Recall from the standard deRham theorems that $\Gamma(M, \mathcal{E}^2)/d\Gamma(M, \mathcal{E}^1) \cong H^2(M,\mathbf{C})$; when the surface M is compact, $H^2(M,\mathbf{C}) \cong \mathbb{C}$ and the deRham correspondence reduces to integrating the differential forms in $\Gamma(M, \mathcal{E}^2)$ over the surface M. In summary then, there is a bilinear Hermitian mapping

$$H^p(M, \mathcal{F}(X)) \otimes H^{2-p}(M, \mathcal{F}(\overline{X}^*)) \longrightarrow \mathbb{C}$$

which associates to the cohomology classes A and B the complex constant

$$(11) \qquad < A,B > = \int_M {}^t\overline{\psi} \wedge \varphi ,$$

over any compact Riemann surface M. The duality theorem is the assertion that this is a dual pairing.

Theorem 20. Let $X \in H^1(M,GL(n,\mathbb{C}))$ be a flat vector bundle over a compact Riemann surface M. The cohomology spaces $H^p(M, \mathcal{F}(X))$ and $H^{2-p}(M, \mathcal{F}(\overline{X}^*))$ are canonically dual to one another, under the pairing (11), for $p = 0,1,2$.

Proof. The proof is just a straightforward adaption of the proof of the Serre duality theorem. On the vector spaces $\Gamma(M, \mathcal{E}^p(X))$ introduce the norms p_n, as defined on page 68. These norms determine the structure of a topological vector space, actually a Frechet space, on $\Gamma(M, \mathcal{E}^p(X)$; and the dual vector space is $\Gamma(M, \mathcal{K}^{2-p}(\overline{X}^*))$, where $\mathcal{K}$ is the sheaf of germs of distributions on the compact Riemann surface M, (recalling Lemma 12). Now consider the sequence of vector spaces

$$(12) \qquad \Gamma(M, \mathcal{E}^{p-1}(X)) \xrightarrow{d} \Gamma(M, \mathcal{E}^p(X)) \xrightarrow{d} \Gamma(M, \mathcal{E}^{p+1}(X)) .$$

The linear mappings d are obviously continuous in the topologies introduced on these spaces, and the image $d\Gamma(M, \mathcal{E}^{p-1}(X)) \subset \Gamma(M, \mathcal{E}^{p}(X))$ is a closed linear subspace; (the latter follows directly from the fact that the quotient space $\ker d/\operatorname{im} d \cong H^p(M, \mathcal{J}(X))$ is finite dimensional, as on page 95 of last year's lectures). Thus in the dual of the exact sequence (12), namely

$$(13) \qquad \Gamma(M, \mathcal{K}^{3-p}(\overline{X}^*)) \xleftarrow{d^*} \Gamma(M, \mathcal{K}^{2-p}(\overline{X}^*)) \xleftarrow{d^*} \Gamma(M, \mathcal{K}^{1-p}(\overline{X}^*)) ,$$

it follows readily that $\ker d^*/\operatorname{im} d^*$ is the dual vector space to $\ker d/\operatorname{im} d \cong H^p(M, \mathcal{J}(X))$. There is an exact sequence of sheaves of germs of distributions over M of the form

$$0 \longrightarrow \mathcal{J}(\overline{X}^*) \longrightarrow \mathcal{K}^0(\overline{X}^*) \xrightarrow{d} \mathcal{K}^1(\overline{X}^*) \xrightarrow{d} \mathcal{K}^2(\overline{X}^*) \longrightarrow 0 ,$$

(see the following Lemma 20); the corresponding exact sequence of sections contains the segment (13), since it follows readily from the definition of the derivative of a distribution that the operator d^* of (13) is just the exterior derivative on distributions. The sheaves of germs of distributions are fine sheaves, hence

$$H^{2-p}(M, \mathcal{J}(\overline{X}^*)) \cong \frac{\Gamma(M, \mathcal{K}_c^{2-p}(\overline{X}^*))}{d\Gamma(M, \mathcal{K}^{1-p}(\overline{X}^*))} = \frac{\ker d^*}{\operatorname{im} d^*} ,$$

which shows that $H^{2-p}(M, \mathcal{J}(\overline{X}^*))$ is the dual space of $H^p(M, \mathcal{J}(X))$. It is an easy matter, which will be left to the readers, to verify that this duality is that given by (11); the proof is then completed, except for the following result.

<u>Lemma 20.</u> Over any Riemann surface M there is an exact sequence of sheaves

$$0 \longrightarrow \mathbb{C} \longrightarrow \mathcal{K}^0 \xrightarrow{d} \mathcal{K}^1 \xrightarrow{d} \mathcal{K}^2 \longrightarrow 0 ,$$

where d is the exterior derivative.

Proof. Let U be a product neighborhood of the origin in the complex plane, so that $U = I \times I$ for some open interval I. It is rather apparent that it suffices to prove the following three assertions; the notation and terminology are as in §6 of last year's lectures.

(i) If $T \in \mathcal{K}_U$ is a distribution such that $\frac{\partial T}{\partial x} = \frac{\partial T}{\partial y} = 0$, then T is a constant.

First select an auxiliary function $h \in {}_o\mathcal{C}_I^\infty$ such that $\int h(t)dt = 1$. For any function $f \in {}_o\mathcal{C}_U^\infty$ set

$$(14) \qquad f(x,y) = f_1(x,y) + h(x) \int f(s,y)ds .$$

The function f_1 is also C^∞ in U, has compact support in U, and moreover satisfies $\int f_1(x,y)dx = 0$; thus there is a C^∞ function $g \in {}_o\mathcal{C}_U^\infty$ such that $f_1 = \partial g/\partial x$, and therefore $T(f_1) = T(\partial g/\partial x) = -(\partial T/\partial x)(g) = 0$. Applying the same idea to the integrand in (14), write

$$f(x,y) = f_1(x,y) + h(x) \int f_2(s,y)ds + h(x)h(y) \iint f(s,t)dsdt$$

where $T(f_2) = 0$. Then

$$\begin{aligned} T(f) &= T(h(x)h(y) \iint f(s,t)dsdt) \\ &= c \cdot \iint f(s,t)dsdt \end{aligned}$$

where $c = T(h(x)h(y))$; but this shows that T is the constant distribution c , as desired.

(ii) For any distribution $T \in \mathcal{K}_U$ there is a distribution $S \in \mathcal{K}_U$ such that $T = \partial S/\partial x$.

Given any function $f \in {}_o\mathcal{C}_U^\infty$, consider again the decomposition (14) as in part (a) above. Setting

$$S(f) = -T(\int_o^x f_1(s,y)ds)$$

yields a distribution $S \in \mathcal{K}_U$. Note that whenever $f = \partial g/\partial x$ for a function $g \in {}_o\mathcal{C}_U^\infty$, then $f = f_1$; so that $(\partial S/\partial x)(g) = -S(\partial g/\partial x) = T(\int_o^x \partial y(s,y)/\partial s \cdot ds) = T(g)$, and $\partial S/\partial x = T$ as desired.

It should be remarked in passing that (14) yields immediately a description of the most general such distribution S . For instance if $T = 0$, or in other words if S is any distribution such that $\partial S/\partial x = 0$, then applying S to (14) it follows that $S(f) = S(f_1) + S(h(x)\int f(s,y)ds) = R_1(\int f(s,y)ds)$, where R_1 is a distribution in $\mathcal{K}_I$.

(iii) If $T_x, T_y \in \mathcal{K}_U$ are distributions such that $\partial T_x/\partial y = \partial T_y/\partial x$, there is a distribution $S \in \mathcal{K}_U$ such that $T_x = \partial S/\partial x$ and $T_y = \partial S/\partial y$.

By part (b) above there will be a distribution $S_1 \in \mathcal{K}_U$ such that $\partial S_1/\partial x = T_x$; indeed, as remarked above, the most general such will be of the form

$$S(f) = S_1(f) + R_1(\int f(s,y)ds)$$

for any distribution $R_1 \in \mathcal{K}_I$. The problem is to choose R_1 so that $\partial S/\partial y = T_y$. Note that

$$\frac{\partial}{\partial x}(T_y - \frac{\partial S_1}{\partial y}) = \frac{\partial T_x}{\partial y} - \frac{\partial}{\partial y}\frac{\partial S_1}{\partial x} = \frac{\partial}{\partial y}(T_x - \frac{\partial S_1}{\partial x}) = 0 ,$$

so that

$$(T_y - \frac{\partial S_1}{\partial y})(f) = R_2(\int f(s,y)ds)$$

for some distribution $R_2 \in \mathcal{K}_I$; thus the condition to be imposed on the distribution R_1 is that for every function $f \in {}_0\mathcal{C}^\infty_U$,

$$0 = (T_y - \frac{\partial S}{\partial y})(f) = (T_y - \frac{\partial S_1}{\partial y})(f) - \frac{\partial R_1}{\partial y}(\int f(s,y)ds)$$

$$= R_2(\int f(s,y)ds) - \frac{\partial R_1}{\partial y}(\int f(s,y)ds) ,$$

or just that $\partial R_1/\partial y = R_2$ in $\mathcal{K}_I$. By part (b) above there always exists such a distribution $R_1 \in \mathcal{K}_I$, and the proof is thereby concluded.

Remarks. A flat vector bundle, and its corresponding sheaf of flat sections, are clearly of a more purely topological than analytical-topological nature. One would expect that there would exist a proof of Theorem 20 of a purely topological sort, as indeed there does. (See Glen E. Bredon, Sheaf Theory (McGraw-Hill, 1967); the discussion there is restricted to flat line bundles, but the extension should be straightforward.) However, since the analytical machinery is at hand, and has been used similarly before, it seemed more reasonable to prove the theorem by that means than to digress further on general sheaf theory.

As a first application, the duality theorem together with Theorem 18 permit the easy calculation of the cohomology group $H^2(M, \mathcal{O}(X))$. If M is a compact Riemann surface and X is a flat vector bundle over M, it follows from Theorem 20 that $H^2(M, \mathcal{O}(X)) \cong H^0(M, \mathcal{O}(\overline{X}^*))$; and therefore, applying Theorem 18,

(15) $\dim_{\mathbf{C}} H^2(M, \mathcal{X}(X)) = r$ where r is the largest integer such that the trivial bundle of rank r is contained as a subbundle of $\overline{X}^*$, for any compact Riemann surface M.

A more interesting application of the duality theorem is to the case $p = 1$; the assertion then is that for any flat vector bundle $X \in H^1(M, GL(n, \mathbf{C}))$ over a compact Riemann surface M, the cohomology groups $H^1(M, \mathcal{O}(X))$ and $H^1(M, \mathcal{O}(\overline{X}^*))$ are canonically dual to one another. Using the isomorphism of Theorem 19, this duality takes the form of a dual pairing

(16) $$H^1(\pi_1(M), \hat{X}) \otimes H^1(\pi_1(M), \hat{\overline{X}}^*) \longrightarrow \mathbf{C} ;$$

and it is of interest to see the explicit form that this pairing takes. Note in particular that when the representation $\hat{X}$ is unitary, so that $\hat{X} = \hat{\overline{X}}^*$, this dual pairing becomes a nonsingular Hermitian-bilinear form on the complex vector space $H^1(\pi_1(M), \hat{X})$; the bundle X is called a unitary flat vector bundle in this case, since it can be defined by a cocycle consisting entirely of unitary matrices. (Recall from last year's lectures that for the special case that X is the trivial line bundle, this is just the intersection matrix of the surface.) The direct relationship between the cohomology group $H^1(\pi_1(M),\hat{X})$ and the deRham group $\Gamma(M, \mathcal{E}^1_c(X)/d\Gamma(M, \mathcal{E}^0(X))$ can be handled most easily by introducing the universal covering surface of M, and transferring the bundles and differential forms to that covering space. The explicit cohomology structure of the surface must eventually be used, of course.

(d) Let $\tilde{M}$ be the universal covering space of the Riemann surface M, and $f\colon \tilde{M} \longrightarrow M$ be the covering mapping. For any flat vector bundle $X \in H^1(M,GL(n,\mathbb{C}))$ over M, with associated sheaf of germs of flat sections $\mathcal{F}(X)$, it is clear that the inverse image sheaf $f^{-1}(\mathcal{F}(X))$, as defined in §3, is a flat sheaf of rank n over the covering surface $\tilde{M}$; hence $f^{-1}(\mathcal{F}(X)) = \mathcal{F}(\tilde{X})$ for some flat vector bundle $\tilde{X} \in H^1(\tilde{M},GL(n,\mathbb{C}))$. Actually the bundle $\tilde{X}$ is the trivial bundle, since $\pi_1(\tilde{M}) = 0$; so that the sheaf $\mathcal{F}(\tilde{X})$ is the trivial sheaf, that is, $\mathcal{F}(\tilde{X}) \cong \mathbb{C}^n \times \tilde{M}$.

Recall that the fundamental group $\pi_1(M)$ can be viewed as a group of transformations of the space $\tilde{M}$ commuting with the mapping $f\colon \tilde{M} \longrightarrow M$; to avoid confusion, this representation should be made quite explicit. Select a base point $p_o \in M$, and consider the fundamental group as the group $\pi_1(M,p_o)$ of homotopy classes of closed paths in M based at p_o; if $\gamma_1,\gamma_2 \in \pi_1(M,p_o)$, the composition $\gamma_1\gamma_2 \in \pi_1(M,p_o)$ is the closed path obtained by traversing first the path γ_1 and then the path γ_2. (The fact that this group is isomorphic to the fundamental group defined by means of open coverings of M, as in last year's lectures, is left as an exercise to the reader.) The universal covering space $\tilde{M}$ can be defined as the space of homotopy classes (with fixed end points) of paths in M based at the point p_o ; the mapping $f: \tilde{M} \longrightarrow M$ is that which assigns to any path its end point. If $\pi \in \tilde{M}$ and $\gamma \in \pi_1(M,p_o)$, then the path $\gamma\pi$, obtained by traversing first the closed path γ then the path π , is also an element of $\tilde{M}$; and the mapping $\pi \longrightarrow \gamma\pi$ exhibits $\pi_1(M,p_o)$ as a group of transformations on the space $\tilde{M}$. The points of $\tilde{M}$ will generally be denoted by $\tilde{p}$ or $\tilde{z}$; and the mapping $\gamma: \tilde{M} \longrightarrow \tilde{M}$ takes the point $\tilde{z} \in \tilde{M}$ to the point denoted by $\gamma \cdot \tilde{z}$. It is clear that $f(\gamma \cdot \tilde{z}) = f(\tilde{z})$ for all $\gamma \in \pi_1(M,p_o)$ and all $\tilde{z} \in \tilde{M}$. Let $\tilde{p}_o$ be the point of $\tilde{M}$ corresponding to the trivial path at p_o , that is to say, to the homotopy class of null-homotopic closed paths at p_o ; this will be called the <u>base point</u> of the covering space $\tilde{M}$. Any closed path $\gamma \in \pi_1(M,p_o)$ is covered by a unique path $\tilde{\gamma} \subset \tilde{M}$ beginning at $\tilde{p}_o$; and it is clear that the end point of the path $\tilde{\gamma}$ is just $\gamma \cdot \tilde{p}_o$. If $\tilde{p}_1 \in \tilde{M}$ is another point such that $f(\tilde{p}_1) = p_o$, there is an element $\sigma \in \pi_1(M,p_o)$ such that $\tilde{p}_1 = \sigma \cdot \tilde{p}_o$; the path $\sigma \cdot \tilde{\gamma} \subset \tilde{M}$ is the unique path beginning at $\tilde{p}_1$ and covering the path γ under the covering mapping f , and the end point of $\sigma \cdot \gamma$ is $\sigma \cdot \gamma \cdot \tilde{p}_o = \sigma\gamma\sigma^{-1} \cdot \tilde{p}_1$.

Since the transformations $\gamma: \tilde{M} \longrightarrow \tilde{M}$ commute with the covering mapping $f: \tilde{M} \longrightarrow M$ for all $\gamma \in \pi_1(M,p_o)$, and $\mathcal{F}(\tilde{X}) = f^{-1}(\mathcal{F}(X))$, it is apparent that these transformations γ extend to sheaf automorphisms $\tilde{\gamma}: \mathcal{F}(\tilde{X}) \longrightarrow \mathcal{F}(\tilde{X})$. For if $\{\tilde{U}_i\}$ are the connected components of the set $f^{-1}(U) \subset \tilde{M}$ for a contractible open subset $U \subset M$, then by definition $\mathcal{F}(\tilde{X})|\tilde{U}_i \cong \mathcal{F}(X)|U_i$; the transformation of $\tilde{M}$ associated to any $\gamma \in \pi_1(M,p_o)$ merely permutes the sets $\tilde{U}_i$, so can be extended to be the same permutation of the restrictions $\mathcal{F}(\tilde{X})|\tilde{U}_i$. In terms of the isomorphism $\mathcal{F}(\tilde{X}) \cong \mathbf{C}^n \times \tilde{M}$, the automorphism of sheaves associated to the group element $\gamma \in \pi_1(M,p_o)$ is a mapping $\tilde{\gamma}: \mathbf{C}^n \times \tilde{M} \longrightarrow \mathbf{C}^n \times \tilde{M}$ which must be of the form

$$\tilde{\gamma} \cdot (V,z) = (\hat{X}(\gamma) \cdot V, \gamma \cdot \tilde{z}) \tag{17}$$

for some matrix $\hat{X}(\gamma) \in GL(n,\mathbf{C})$, where $V \in \mathbf{C}^n$ is a column vector and $\tilde{z} \in \tilde{M}$. The mapping $\gamma \longrightarrow \hat{X}(\gamma)$ is clearly a group homomorphism $\hat{X}: \pi_1(M,p_o) \longrightarrow GL(n,\mathbf{C})$; note that the representation $\hat{X}$ is only determined up to an inner automorphism of $GL(n,\mathbf{C})$, since the isomorphism $\mathcal{F}(\tilde{X}) \cong \mathbf{C}^n \times \tilde{M}$ is of course not unique.

Lemma 21. The homomorphism $\hat{X}$ is the characteristic representation of the flat vector bundle X.

Proof. This is a straightforward matter of examining more explicitly the above construction. Let $\mathcal{U} = \{U_\alpha\}$ be an open covering of M such that the sets U_α and all their intersections are contractible; thus $\mathcal{U}$ is a Leray covering for flat sheaves, and $\pi_1(M,p_o) \cong \pi_1(\mathcal{U},U_o)$ for fixed base points $p_o \in U_o$. For any set $U_\alpha \in \mathcal{U}$ the inverse image $f^{-1}(U_\alpha)$ in the covering space $\tilde{M}$ is the union of countably many open components $\tilde{U}_{\alpha i}$, and these components form an open covering $\tilde{\mathcal{U}} = \{\tilde{U}_{\alpha i}\}$ of $\tilde{M}$; let $\tilde{U}_{oi_o}$ be that component of $f^{-1}(U_o)$ containing the base point $\tilde{p}_o$ of the covering. The flat vector bundle X can be represented by a cocycle $(X_{\alpha\beta}) \in Z^1(\mathcal{U},GL(n,\mathbf{C}))$; this corresponds to a choice of an isomorphism $\mathcal{F}(X)|U_\alpha \cong \mathbf{C}^n \times U_\alpha$ for each open set U_α. Selecting for the induced sheaf $\mathcal{F}(\tilde{X})$ the corresponding isomorphism $\mathcal{F}(\tilde{X})|\tilde{U}_{\alpha i} \cong \mathbf{C}^n \times \tilde{U}_{\alpha i}$, it is clear that the bundle $\tilde{X}$ is represented by the cocycle $(\tilde{X}_{\alpha i,\beta j}) \in Z^1(\tilde{\mathcal{U}}, GL(n,\mathbf{C}))$, where $\tilde{X}_{\alpha i,\beta j} = X_{\alpha\beta}$ whenever $\tilde{U}_{\alpha i} \cap \tilde{U}_{\beta j} \neq \emptyset$. Since the bundle $\tilde{X}$ is trivial, there are constant matrices $\tilde{C}_{\alpha i} \in GL(n,\mathbf{C})$ such that $\tilde{C}_{\alpha i}\tilde{X}_{\alpha i,\beta j}\tilde{C}_{\beta j}^{-1} = I$ whenever $\tilde{U}_{\alpha i} \cap \tilde{U}_{\beta j} \neq \emptyset$; the isomorphisms $\tilde{C}_{i\alpha}: \mathbf{C}^n \times \tilde{U}_{\alpha i} \longrightarrow \mathbf{C}^n \times \tilde{U}_{\alpha i}$ exhibit explicitly the reduction of the sheaf $\mathcal{F}(\tilde{X})$ to the trivial product sheaf $\mathbf{C}^n \times \tilde{M}$. There is no loss of generality in supposing that $\tilde{C}_{0i_0} = I$; all the matrices $\tilde{C}_{\alpha i}$ are then uniquely determined. Now for any closed chain $\gamma = (U_{\alpha_0}, U_{\alpha_1}, \ldots, U_{\alpha_q}) \in \pi_1(\mathcal{U}, U_0)$ there is a unique chain $\tilde{\gamma} = (\tilde{U}_{\alpha_0 i_0}, \tilde{U}_{\alpha_1 i_1}, \ldots, \tilde{U}_{\alpha_q i_q})$ in $\tilde{\mathcal{U}}$ based at $\tilde{U}_{\alpha_0 i_0} = \tilde{U}_{0 i_0}$ and covering γ under the mapping f. The chain $\tilde{\gamma}$ need not be closed, but $f(\tilde{U}_{\alpha_q i_q}) = U_0$; and the transformation of $\tilde{M}$ associated to γ is the covering translation taking $\tilde{U}_{\alpha_0 i_0}$ to $\tilde{U}_{\alpha_q i_q}$. The sheaf automorphism associated to γ is the identity mapping on the factor $\mathbf{C}^n$, when the bundle $\tilde{X}$ is viewed as defined by the cocycle $(\tilde{X}_{\alpha i,\beta j})$; so applying the isomorphisms $\tilde{C}_{\alpha i}$ to reduce the sheaf $\mathcal{F}(\tilde{X})$ to the trivial sheaf, the automorphism associated to γ has on the factor $\mathbf{C}^n$ the form $\hat{X}(\gamma) = \tilde{C}_{\alpha_q i_q}\tilde{C}_{\alpha_0 i_0}^{-1} = \tilde{C}_{\alpha_q i_q}$. However, since $\tilde{C}_{\alpha_{j+1} i_{j+1}} = \tilde{C}_{\alpha_j i_j}\tilde{X}_{\alpha_j i_j, \alpha_{j+1} i_{j+1}} = \tilde{C}_{\alpha_j i_j} X_{\alpha_j \alpha_{j+1}}$, it it follows that

(18) $$\hat{X}(\gamma) = \tilde{C}_{\alpha_q i_q} = X_{\alpha_0\alpha_1} X_{\alpha_1\alpha_2} \cdots X_{\alpha_{q-1}\alpha_q} ;$$

but this is just the characteristic representation of the bundle X, recalling the construction in last year's lectures, and the proof is thereby concluded.

Now consider a cohomology class $A \in H^1(M, \mathcal{F}(X))$. Under the deRham isomorphism (10) this cohomology class can be represented by a closed differential form $\varphi \in \Gamma(M, \mathcal{E}^1(X))$; and φ lifts to a closed differential form $f^*\varphi \in \Gamma(\tilde{M}, \mathcal{E}^1(\tilde{X}))$ under the induced mapping associated to the covering $f: \tilde{M} \longrightarrow M$. The form $f^*\varphi$ is clearly invariant under the automorphism $\gamma: \Gamma(\tilde{M}, \mathcal{E}^1(\tilde{X})) \longrightarrow \Gamma(\tilde{M}, \mathcal{E}^1(\tilde{X}))$ associated to a covering translation $\gamma \in \pi_1(M,p_0)$. Since the bundle $\tilde{X}$ is trivial, under the isomorphism $\mathcal{E}^1(\tilde{X}) \cong (\mathcal{E}^1)^n$ the differential form $f^*\varphi$ can be viewed as an n-tuple $\tilde{\varphi}$ of differential forms on the manifold $\tilde{M}$ in the ordinary sense, that is, as an element $\tilde{\varphi} \in \Gamma(\tilde{M}, (\mathcal{E}^1)^n)$; the form $\tilde{\varphi}$ is clearly closed, and satisfies

$$\tilde{\varphi}(\gamma \cdot \tilde{z}) = \hat{X}(\gamma) \cdot \tilde{\varphi}(\tilde{z}) \tag{19}$$

for every covering translation $\gamma \in \pi_1(M,p_0)$. The cohomology class A can also be represented by a cohomology class $\hat{A} \in H^1(\pi_1(M), \hat{X})$ under the isomorphism of Theorem 19; the cohomology class $\hat{A}$ and the differential form $\tilde{\varphi}$ are related quite simply, as follows.

<u>Lemma 22.</u> Letting $\tilde{p}_0 \in \tilde{M}$ be the base point of the covering space $\tilde{M}$, and for each loop $\gamma \in \pi_1(M,p_0)$ letting $\tilde{\gamma}$ be the path in $\tilde{M}$ covering γ and based at the point $\tilde{p}_0$, the cohomology class $\hat{A} \in H^1(\pi_1(M), \hat{X})$ is represented by the cocycle $(\hat{A}'_\gamma) \in Z^1(\pi_1(M), \hat{X})$, where

$$\hat{A}'_\gamma = -\hat{X}(\gamma)^{-1} \int_{\tilde{\gamma}} \tilde{\varphi} .$$

Proof. Select an open covering $\mathcal{U} = \{U_\alpha\}$ of the surface M as in the proof of Lemma 21, and continue with the constructions and notation as in that proof. Let $(A_{\alpha\beta}) \in Z^1(\mathcal{U}, \mathcal{F}(X))$ be a cocycle representing the cohomology class A, and select C^∞ vector-valued functions F_α in the various sets U_α such that $A_{\alpha\beta} = F_\beta - X_{\alpha\beta}^{-1} F_\alpha$ in $U_\alpha \cap U_\beta$; the differential form φ representing A under the deRham isomorphism (10) can be taken to be $\varphi_\alpha = dF_\alpha$ in U_α. Reducing the bundle $\tilde{X}$ to the trivial bundle as in the proof of Lemma 21, the differential form $\tilde{\varphi}$ is clearly given by $\tilde{\varphi}_{\alpha i} = \tilde{C}_{\alpha i}\varphi_\alpha = \tilde{C}_{\alpha i} dF_\alpha$ in each set $\tilde{U}_{\alpha i}$. For any closed path $\gamma \in \pi_1(M, p_o)$, the lifted path $\tilde{\gamma}$ can be covered by a chain $(\tilde{U}_{\alpha_o i_o}, \tilde{U}_{\alpha_1 i_1}, \ldots, \tilde{U}_{\alpha_q i_q})$ of the covering $\tilde{\mathcal{U}}$, where $\tilde{U}_{\alpha_o i_o} = \tilde{U}_{o i_o}$ and $f(\tilde{\gamma}) = (U_{\alpha_o}, U_{\alpha_1}, \ldots, U_{\alpha_q}) \in \pi_1(\mathcal{U}, U_o)$. Then the path $\tilde{\gamma}$ can be decomposed into non-overlapping segments $\tilde{\gamma}_j$ for $j = 0, 1, \ldots, q$, such that $\tilde{\gamma}_j$ lies entirely within the set $\tilde{U}_{\alpha_j i_j}$; the end points of the segment $\tilde{\gamma}_j$ will be denoted by $\tilde{p}_j$ and $\tilde{p}_{j+1}$, so that $\tilde{p}_o$ and $\tilde{p}_{q+1}$ are the end points of the full path $\tilde{\gamma}$, and $\tilde{p}_j \in \tilde{U}_{\alpha_{j-1} i_{j-1}} \cap \tilde{U}_{\alpha_j i_j}$ for $j = 1, 2, \ldots, q$. Now

$$\int_{\tilde{\gamma}} \tilde{\varphi} = \sum_{j=o}^{q} \int_{\tilde{\gamma}_j} \tilde{\varphi}_{\alpha_j i_j} = \sum_{j=o}^{q} \int_{\tilde{\gamma}_j} \tilde{C}_{\alpha_j i_j}\, dF_{\alpha_j}$$

$$= \sum_{j=o}^{q} \tilde{C}_{\alpha_j i_j} [F_{\alpha_j}(p_{j+1}) - F_{\alpha_j}(p_j)] , \quad \text{for } p_j = f(\tilde{p}_j) ,$$

$$= \tilde{C}_{\alpha_q i_q} F_{\alpha_q}(p_{q+1}) - \tilde{C}_{\alpha_o i_o} F_{\alpha_o}(p_o) +$$

$$+ \sum_{j=1}^{q} [\tilde{C}_{\alpha_{j-1} i_{j-1}} F_{\alpha_{j-1}}(p_j) - \tilde{C}_{\alpha_j i_j} F_{\alpha_j}(p_j)] .$$

Here of course $F_{\alpha_q}(p_{q+1}) = F_{\alpha_o}(p_o)$, and from the definition of these functions it follows that

$$F_{\alpha_{j-1}}(p_j) = \chi_{\alpha_{j-1}\alpha_j}(F_{\alpha_j}(p_j) - A_{\alpha_{j-1}\alpha_j}) .$$

Furthermore, as in the proof of Lemma 21, the constants $\tilde{C}_{\alpha_j i_j}$ are so chosen that $\tilde{C}_{\alpha_o i_o} = I$ and $\tilde{C}_{\alpha_{j-1} i_{j-1}} \tilde{\chi}_{\alpha_{j-1} i_{j-1}, \alpha_j i_j} = \tilde{C}_{\alpha_{j-1} i_{j-1}} \chi_{\alpha_{j-1}\alpha_j} = \tilde{C}_{\alpha_j i_j}$; thus $\tilde{C}_{\alpha_j i_j} = \chi_{\alpha_o\alpha_1}\chi_{\alpha_1\alpha_2} \cdots \chi_{\alpha_{j-1}\alpha_j}$ and $\tilde{C}_{\alpha_q i_q} = \hat{\chi}(\gamma)$. It then follows that

$$\int_{\tilde{\gamma}} \tilde{\varphi} = \hat{\chi}(\gamma) \cdot F_{\alpha_o}(p_o) - F_{\alpha_o}(p_o) - \sum_{j=1}^{q} \tilde{C}_{\alpha_j i_j} A_{\alpha_{j-1}\alpha_j}$$

so that

$$\begin{aligned} -\hat{\chi}(\gamma)^{-1} \int_{\tilde{\gamma}} \tilde{\varphi} &= -F_{\alpha_o}(p_o) + \hat{\chi}(\gamma)^{-1} \cdot F_{\alpha_o}(p_o) + \\ &\quad + \sum_{j=1}^{q} (\chi_{\alpha_j\alpha_{j+1}} \cdots \chi_{\alpha_{q-1}\alpha_q})^{-1} A_{\alpha_{j-1}\alpha_j} \\ &= -F_{\alpha_o}(p_o) + \hat{\chi}(\gamma)^{-1} \cdot F_{\alpha_o}(p_o) + \hat{A}_\gamma , \end{aligned}$$

where $\hat{A}_\gamma$ is as defined in the proof of Theorem 19, and represents the cohomology class $\hat{A}$; thus $-\hat{\chi}(\gamma)^{-1} \int_{\tilde{\gamma}} \tilde{\varphi}$ is cohomologous to $\hat{A}_\gamma$, hence represents the cohomology class $\hat{A}$ as well, which concludes the proof.

Remarks. Recall that the path $\tilde{\gamma}$ was so chosen that it begins at $\tilde{p}_o \in \tilde{M}$ and ends at $\gamma \cdot \tilde{p}_o \in \tilde{M}$. Since the differential form $\tilde{\varphi}$ is closed and the space $\tilde{M}$ is simply-connected, the integral $\int_{\tilde{\gamma}} \tilde{\varphi}$ is unchanged if $\tilde{\gamma}$ is replaced by any other path from $\tilde{p}_o$ to $\gamma \cdot \tilde{p}_o$; thus Lemma 22 can be restated as the assertion that the cohomology class $\hat{A}$ is represented by the cocycle $(\hat{A}_\gamma)$ where

$$(20) \qquad \hat{A}_\gamma = -\hat{X}(\gamma)^{-1} \int_{\tilde{p}_o}^{\gamma\cdot\tilde{p}_o} \tilde{\varphi} \ .$$

More generally, introduce the C^∞ function $F\colon \tilde{M} \longrightarrow \mathbf{C}^n$ given by

$$(21) \qquad F(\tilde{z}) = \int_{\tilde{p}_o}^{\tilde{z}} \tilde{\varphi} \ ,$$

noting as above that this is well-defined. Thus $\hat{A}_\gamma = -\hat{X}(\gamma)^{-1} \cdot F(\gamma\cdot\tilde{p}_o)$; and further, for every $\tilde{z} \in \tilde{M}$, recalling (19),

$$F(\gamma\cdot\tilde{z}) = \int_{\tilde{p}_o}^{\gamma\cdot\tilde{p}_o} \tilde{\varphi} + \int_{\gamma\cdot\tilde{p}_o}^{\gamma\cdot\tilde{z}} \tilde{\varphi} = \int_{\tilde{p}_o}^{\gamma\cdot\tilde{p}_o} \tilde{\varphi} + \hat{X}(\gamma) \cdot \int_{\tilde{p}_o}^{\tilde{z}} \tilde{\varphi} = \hat{X}(\gamma)[F(\tilde{z}) - \hat{A}_\gamma] \ ,$$

so that

$$(22) \qquad \hat{A}_\gamma = F(\tilde{z}) - \hat{X}(\gamma)^{-1} \cdot F(\gamma\cdot\tilde{z}) \ .$$

It is apparent from this that the choice of the base point $\tilde{p}_o \in \tilde{M}$ is completely immaterial. For if $\tilde{p}_1$ is any other point of $\tilde{M}$ and $G(z) = \int_{\tilde{p}_1}^{\tilde{z}} \tilde{\varphi}$, then $G(\tilde{z}) = F(\tilde{z}) + C$ where $C = \int_{\tilde{p}_1}^{\tilde{p}_o} \tilde{\varphi}$; and

$$\hat{B}_\gamma = G(\tilde{z}) - \hat{X}(\gamma)^{-1} \cdot G(\gamma\cdot\tilde{z}) = \hat{A}_\gamma + (C - \hat{X}(\gamma)^{-1} \cdot C) \ ,$$

so the cocycles $(\hat{A}_\gamma)$ and $(\hat{B}_\gamma)$ are cohomologous and thus represent the same element in $H^1(\pi_1(M),\hat{X})$.

As for terminology, the cohomology class $A = (\hat{A}_\gamma)$ will be called the <u>period class</u> of the differential form $\tilde{\varphi}$; this is the cohomology class defined by the cocycle $\hat{A}_\gamma$ given by (21) and (22), where $\tilde{\varphi}$ is any closed differential form on $\tilde{M}$ satisfying (19). When X is the trivial bundle of rank 1, the period class is just the homomorphism $\hat{A}\colon \pi_1(M) \longrightarrow \mathbf{C}$ determined by the periods of the differential form $\tilde{\varphi}$, whence the terminology.

(e) To determine the explicit form of the dual pairing (16), it is necessary to use rather explicitly the topological structure of the surface. If M is a compact Riemann surface of genus g and $p_0 \in M$ is a fixed but arbitrary base point, let $\sigma_j, \tau_j \in \pi_1(M, p_0)$, $j = 1, \ldots, g$, be a standard set of return cuts on the surface. Thus σ_j, τ_j are closed loops on M, which are disjoint except for the point p_0 and which dissect M into a simply-connected surface, in the sense that $\Delta = M - \bigcup_j(\sigma_j \cup \tau_j)$ is simply connected. (See Seifert-Threlfall, Lehrbuch der Topologie, for instance.) The loops will be assumed to be labeled in the order shown on the following diagram.

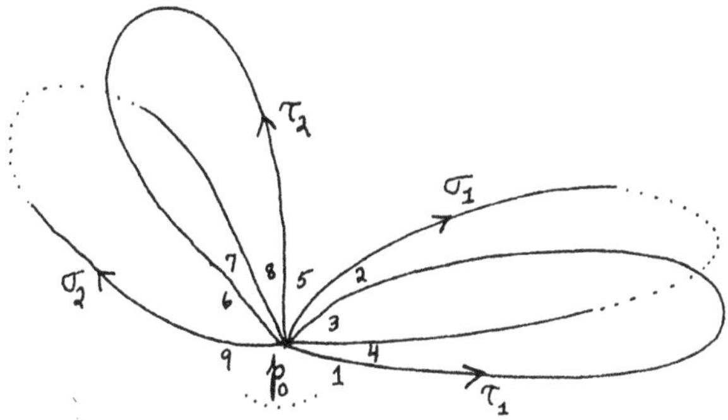

Upon tracing along these loops in order, as indicated by the numbers on the preceding diagram, it is apparent that Δ can be viewed as a polygon of $4g$ sides, each loop σ_j, τ_j determining two separate sides; and the surface M can be recovered from the (closed) polygon Δ by identifying appropriate pairs of opposite sides. Letting $\tilde{p}_0 \in \tilde{M}$ be the base point of the covering, where $f: \tilde{M} \longrightarrow M$ is the universal covering space, the polygon $\Delta \subset M$ can be lifted to a unique polygon $\tilde{\Delta} \subset \tilde{M}$, where $\tilde{\Delta}$ is bounded by a loop beginning at $\tilde{p}_o$ and following along the covering of the loops σ_j, τ_j in the order indicated. Thus $\tilde{\Delta} \subset \tilde{M}$ is an open 2-cell, and its boundary consists of $4g$ paths covering the loops σ_j, τ_j ; this is clearer upon considering the following diagram.

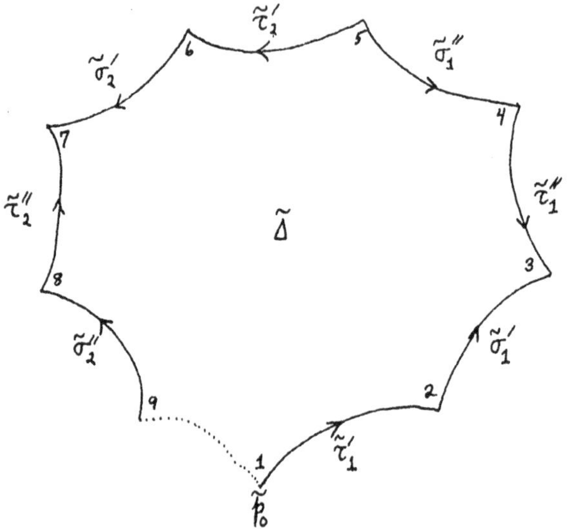

Here $\tilde{\tau}_j', \tilde{\tau}_j''$ are paths in $\tilde{M}$ which cover the loop $\tau_j \in \pi_1(M, p_o)$, with the orientations as indicated on the diagram. It is clear that tracing along the boundary of $\tilde{\Delta}$ and projecting that path into M, determines a loop in $\pi_1(M, p_o)$ which is homotopic to zero; and thus one secures the relation

$$\tau_1\sigma_1\tau_1^{-1}\sigma_1^{-1} \cdots \tau_g\sigma_g\tau_g^{-1}\sigma_g^{-1} = 1$$

in the group $\pi_1(M, p_o)$, which is of course equivalent to the relation (5).

Let $\tilde{\tau}_j, \tilde{\sigma}_j$ be paths in $\tilde{M}$ beginning at $\tilde{p}_o$ and covering the loops τ_j, σ_j respectively; as noted earlier, the endpoint of the path $\tilde{\tau}_j$ is the point $\tau_j \cdot \tilde{p}_o$, and the endpoint of the path $\tilde{\sigma}_j$ is the point $\sigma_j \cdot \tilde{p}_o$, viewing $\pi_1(M,p_o)$ as acting as a group of transformations on $\tilde{M}$. Note then that $\tilde{\tau}_1' = \tilde{\tau}_1$, so ends at the point $\tau_1\tilde{p}_o$; and since $\tilde{\sigma}_1'$ begins at $\tau_1\tilde{p}_o$, necessarily $\tilde{\sigma}_1' = \tau_1 \cdot \tilde{\sigma}_1$ and $\tilde{\sigma}_1'$ ends at $\tau_1\sigma_1 \cdot \tilde{p}_o$; and since $\tilde{\tau}_1''$ ends at the point $\tau_1\sigma_1 \cdot \tilde{p}_o$, necessarily $\tilde{\tau}_1'' = \tau_1\sigma_1\tau_1^{-1} \cdot \tilde{\tau}_1$ and $\tilde{\tau}_1''$ begins at the point $\tau_1\sigma_1\tau_1^{-1} \cdot \tilde{p}_o$. Continuing in this way around the boundary of $\tilde{\Delta}$, all the arcs and vertices can readily be identified as suitable translations of the basic paths $\tilde{\tau}_j, \tilde{\sigma}_j$ and the point $\tilde{p}_o$. In general, let

$$(23) \qquad \lambda_j = [\tau_1,\sigma_1] \cdots [\tau_j,\sigma_j] \in \pi_1(M,p_o) ,$$

where $[\tau,\sigma] = \tau\sigma\tau^{-1}\sigma^{-1}$ as before. It is then readily verified that

$$(24) \qquad \begin{cases} \tilde{\tau}_j' = \lambda_{j-1} \cdot \tilde{\tau}_j ; & \tilde{\tau}_j'' = \lambda_{j-1}\tau_j\sigma_j\tau_j^{-1} \cdot \tilde{\tau}_j ; \\ \tilde{\sigma}_j' = \lambda_{j-1}\tau_j \cdot \tilde{\sigma}_j ; & \tilde{\sigma}_j'' = \lambda_j\tilde{\sigma}_j , \end{cases}$$

for $j = 1,\ldots,g$. With this notation, the dual pairing (16) takes the following form.

Theorem 21. Let M be a compact Riemann surface of genus g, and $X \in H^1(M,GL(n,\mathbf{C}))$ be a flat vector bundle over M; and consider cohomology classes $A \in H^1(M, \mathcal{F}(X))$ and $B \in H^1(M, \mathcal{F}(X^*))$. In terms of representative cocycles $(A_\gamma) \in Z^1(\pi_1(M),\hat{X})$ and $(B_\gamma) \in Z^1(\pi_1(M), \hat{X}^*)$, the dual pairing associates to these cohomology classes the value

$$< A,B > = \sum_{j=1}^{g} [{}^t\overline{B}_{\tau_j} \hat{X}(\tau_j)^{-1} A_{\sigma_j} - {}^t\overline{B}_{\sigma_j} \hat{X}(\sigma_j)^{-1} A_{\tau_j}]$$

$$+ \sum_{j=1}^{g} [{}^t\overline{B}_{\tau_j} \hat{X}(\sigma_j\tau_j)^{-1}(A_{\lambda_j} - \hat{X}(\sigma_j)A_{\lambda_{j-1}}) - {}^t\overline{B}_{\sigma_j} \hat{X}(\tau_j\sigma_j)^{-1}(A_{\lambda_{j-1}} - \hat{X}(\tau_j)A_{\lambda_j})].$$

Proof. Let $\varphi \in \Gamma(M, \mathcal{E}^1_c(X))$ be a differential form representing the cohomology class A under the deRham isomorphism (10), and let $\tilde{\varphi}$ be the induced differential form on $\tilde{M}$; thus $\tilde{\varphi}(\gamma \cdot \tilde{z}) = \hat{X}(\gamma) \cdot \tilde{\varphi}(\tilde{z})$ for all $\gamma \in \pi_1(M, p_0)$, and the cocycle $A_\gamma \in Z^1(\pi_1(M), \hat{X})$ representing A is given explicitly by $A_\gamma = F(\tilde{z}) - \hat{X}(\gamma)^{-1} \cdot F(\gamma \cdot \tilde{z})$ where $dF = \tilde{\varphi}$ and $F(\tilde{p}_0) = 0$. Similarly, let $\psi \in \Gamma(M, \mathcal{E}^1_c(\overline{X}^*))$ represent the cohomology class B , and $\tilde{\psi}$ be its induced differential form on $\tilde{M}$; thus $\tilde{\psi}(\gamma \cdot \tilde{z}) = \hat{\overline{X}}^*(\gamma) \cdot \tilde{\psi}(\tilde{z})$ and $B_\gamma = G(\tilde{z}) - \hat{\overline{X}}^*(\gamma)^{-1} \cdot G(\gamma \cdot \tilde{z})$ where $dG = \tilde{\psi}$ and $G(\tilde{p}_0) = 0$. The dual pairing of the cohomology classes A and B is given by

$$< A,B > = \int_M {}^t\overline{\psi} \wedge \varphi = \int_{\Delta} {}^t\tilde{\overline{\psi}} \wedge \tilde{\varphi} = \int_{\Delta} {}^t d\overline{G} \wedge dF .$$

Applying Stokes' theorem and (24),

$$< A,B > = -\int_{\partial\Delta} {}^t d\overline{G} \cdot F = -\sum_{j=1}^{g} \int_{\tilde{\tau}'_j + \tilde{\sigma}'_j - \tilde{\tau}''_j - \tilde{\sigma}''_j} {}^t d\overline{G} \cdot F$$

$$= -\sum_{j=1}^{g} \int_{z\in\tilde{\tau}_j} [{}^t d\overline{G}(\lambda_{j-1} \cdot \tilde{z}) \cdot F(\lambda_{j-1} \cdot \tilde{z}) - {}^t d\overline{G}(\lambda_{j-1}\tau_j\sigma_j\tau_j^{-1} \cdot \tilde{z}) \cdot F(\lambda_{j-1}\tau_j\sigma_j\tau_j^{-1} \cdot \tilde{z})]$$

$$- \sum_{j=1}^{g} \int_{\tilde{z}\in\tilde{\sigma}_j} [{}^t d\overline{G}(\lambda_{j-1}\tau_j \cdot \tilde{z}) \cdot F(\lambda_{j-1}\tau_j \cdot \tilde{z}) - {}^t d\overline{G}(\lambda_j \cdot \tilde{z}) \cdot F(\lambda_j \cdot \tilde{z})] ;$$

and then, recalling the above functional equations for the functions F and G ,

$$< A,B > = -\sum_{j=1}^{g} \int_{\tilde{z}\in\tilde{\tau}_j} [{}^t d\overline{G}(\tilde{z})\cdot(F(\tilde{z})-A_{\lambda_{j-1}}) - {}^t d\overline{G}(\tilde{z})\cdot(F(\tilde{z})-A_{\lambda_{j-1}\tau_j\sigma_j\tau_j^{-1}})]$$

$$-\sum_{j=1}^{g} \int_{\tilde{z}\in\tilde{\sigma}_j} [{}^t d\overline{G}(\tilde{z})\cdot(F(\tilde{z}) - A_{\lambda_{j-1}\tau_j}) - {}^t d\overline{G}(\tilde{z}) \cdot (F(\tilde{z}) - A_{\lambda_j})] .$$

Since $\int_{\tilde{\tau}} {}^t d\overline{G} = {}^t\overline{G}(\tau\cdot\tilde{p}_o) - {}^t\overline{G}(\tilde{p}_o) = -\, {}^t\overline{B}_\tau \cdot \hat{X}(\tau)^{-1}$, the above further reduces to

$$< A,B > = -\sum_{j=1}^{g} {}^t\overline{B}_{\tau_j} \hat{X}(\tau_j)^{-1} [A_{\lambda_{j-1}} - A_{\lambda_{j-1}\tau_j\sigma_j\tau_j^{-1}}]$$

$$-\sum_{j=1}^{g} {}^t\overline{B}_{\sigma_j} \hat{X}(\sigma_j)^{-1} [A_{\lambda_{j-1}\tau_j} - A_{\lambda_j}]$$

$$= -\sum_{j=1}^{g} {}^t\overline{B}_{\tau_j} \hat{X}(\tau_j)^{-1} [A_{\lambda_{j-1}} - \hat{X}(\sigma_j)^{-1}A_{\lambda_j} - A_{\sigma_j}]$$

$$-\sum_{j=1}^{g} {}^t\overline{B}_{\sigma_j} \hat{X}(\sigma_j)^{-1} [\hat{X}(\tau_j)^{-1} A_{\lambda_{j-1}} + A_{\tau_j} - A_{\lambda_j}]$$

$$= \sum_{j=1}^{g} [{}^t\overline{B}_{\tau_j} \hat{X}(\tau_j)^{-1} A_{\sigma_j} - {}^t\overline{B}_{\sigma_j} \hat{X}(\sigma_j)^{-1} A_{\tau_j}]$$

$$+\sum_{j=1}^{g} {}^t\overline{B}_{\tau_j} \hat{X}(\sigma_j\tau_j)^{-1} [A_{\lambda_j} - \hat{X}(\sigma_j)A_{\lambda_{j-1}}]$$

$$-\sum_{j=1}^{g} {}^t\overline{B}_{\sigma_j} \hat{X}(\tau_j\sigma_j)^{-1} [A_{\lambda_{j-1}} - \hat{X}(\tau_j)A_{\lambda_j}] ,$$

which is the desired result.

<u>Remarks.</u> The formula for the dual pairing as given in Theorem 21 can be simplified somewhat by using the formal properties of cocycles. Recall that $A_{\gamma_1\gamma_2} = \hat{X}(\gamma_2^{-1}) \cdot A_{\gamma_1} + A_{\gamma_2}$ and ${}^t\overline{B}_{\gamma_1\gamma_2} = {}^t\overline{B}_{\gamma_1} \cdot \hat{X}(\gamma_2) + {}^t\overline{B}_{\gamma_2}$ for all $\gamma_1,\gamma_2 \in \pi_1(M)$; thus $A_{\gamma^{-1}} = -\hat{X}(\gamma) \cdot A_\gamma$ and ${}^t\overline{B}_{\gamma^{-1}} = -{}^t\overline{B}_\gamma \cdot \hat{X}(\gamma^{-1})$. Recall further that

$\lambda_j = \lambda_{j-1} \cdot [\tau_j, \sigma_j]$, hence that $A_{\lambda_j} = \hat{X}([\sigma_j, \tau_j]) \cdot A_{\lambda_{j-1}} + A_{[\sigma_j, \tau_j]}$. Thus

$$< A,B > = \sum_{j=1}^{g} ({}^t\overline{B}_{\sigma_j^{-1}} \cdot A_{\tau_j} - {}^t\overline{B}_{\tau_j^{-1}} \cdot A_{\sigma_j})$$

$$= \sum_{j=1}^{g} \left\{ {}^t\overline{B}_{\tau_j} \left[\hat{X}(\sigma_j^{-1}\tau_j^{-1}) \cdot A_{\lambda_{j-1}} - \hat{X}(\tau_j^{-1}) \cdot A_{\lambda_{j-1}} + \hat{X}(\tau_j^{-1}\sigma_j^{-1}) \cdot A_{[\sigma_j,\tau_j]} \right] \right.$$

$$\left. - {}^t\overline{B}_{\sigma_j} \left[\hat{X}(\sigma_j^{-1}\tau_j^{-1}) \cdot A_{\lambda_{j-1}} - \hat{X}(\tau_j\sigma_j^{-1}\tau_j^{-1}) \cdot A_{\lambda_{j-1}} - \hat{X}(\sigma_j^{-1}) \cdot A_{[\sigma_j,\tau_j]} \right] \right\}$$

$$= \sum_{j=1}^{g} \left\{ \left[{}^t\overline{B}_{\tau_j} + {}^t\overline{B}_{\sigma_j} \cdot \hat{X}(\tau_j) \right] \hat{X}(\tau_j^{-1}\sigma_j^{-1}) \, A_{[\sigma_j,\tau_j]} \right.$$

$$\left. + \left[{}^t\overline{B}_{\tau_j} (I - \hat{X}(\sigma_j)) - {}^t\overline{B}_{\sigma_j} (I - \hat{X}(\tau_j)) \right] \hat{X}(\sigma_j^{-1}\tau_j^{-1}) \, A_{\lambda_{j-1}} \right\}$$

$$= \sum_{j=1}^{g} \left\{ {}^t\overline{B}_{\sigma_j\tau_j} \, \hat{X}(\tau_j^{-1}\sigma_j^{-1}) \, A_{[\sigma_j,\tau_j]} + {}^t\overline{B}_{[\sigma_j,\tau_j]} \, A_{\lambda_{j-1}} \right\} ;$$

so

$$(25) \quad < A,B > = \sum_{j=1}^{g} {}^t\overline{B}_{\sigma_j^{-1}} \cdot A_{\tau_j} - {}^t\overline{B}_{\tau_j^{-1}} \cdot A_{\sigma_j} - {}^t\overline{B}_{(\sigma_j\tau_j)^{-1}} \cdot A_{[\sigma_j,\tau_j]} + {}^t\overline{B}_{[\sigma_j,\tau_j]} \cdot A_{\lambda_{j-1}}$$

§8. Flat sheaves: analytic aspects

(a) Consider a flat vector bundle $X \in H^1(M, GL(n,\mathbf{C}))$ over a compact Riemann surface M of genus g. The complex analytic version of the deRham sequence considered in §7(c) above is the following exact sequence of sheaves:

$$(1) \qquad 0 \longrightarrow \mathcal{F}(X) \longrightarrow \mathcal{O}(X) \xrightarrow{d} \mathcal{O}^{1,0}(X) \longrightarrow 0 .$$

The associated exact cohomology sequence over M, which in a sense is the basis for the complex analytic properties of flat sheaves, has the form

$$(2) \quad 0 \longrightarrow \Gamma(M, \mathcal{F}(X)) \longrightarrow \Gamma(M, \mathcal{O}(X)) \xrightarrow{d} \Gamma(M, \mathcal{O}^{1,0}(X)) \longrightarrow$$
$$\longrightarrow H^1(M, \mathcal{F}(X)) \longrightarrow H^1(M, \mathcal{O}(X) \xrightarrow{d} H^1(M, \mathcal{O}^{1,0}(X)) \longrightarrow$$
$$\longrightarrow H^2(M, \mathcal{F}(X)) \longrightarrow 0 ,$$

since $H^2(M, \mathcal{O}(X)) = 0$.

Lemma 23. Over a compact Riemann surface M, the kernel of the homomorphism $d\colon H^1(M, \mathcal{O}(X)) \longrightarrow H^1(M, \mathcal{O}^{1,0}(X))$ in the exact sequence (2) is canonically isomorphic to the space

$$\left[\frac{\Gamma(M, \mathcal{O}^{1,0}(X^*))}{d\Gamma(M, \mathcal{O}(X^*))}\right]^* ,$$

where $[\cdot]^*$ denotes the dual complex vector space.

Proof. Considering the exact sequence (2) and the corresponding exact sequence associated to the dual flat vector bundle $X^* \in H^1(M, GL(n,\mathbf{C}))$, the Serre duality theorem exhibits the dual pairings indicated by $\longleftrightarrow$ in the following diagram of exact sequences.

$$\begin{array}{ccccccccc} 0 & \longrightarrow & K & \longrightarrow & H^1(M, \mathcal{O}(X)) & \xrightarrow{d} & H^1(M, \mathcal{O}^{1,0}(X)) & \longrightarrow & \cdots \\ & & & & \updownarrow & & \updownarrow & & \\ 0 & \longleftarrow & L & \longleftarrow & \Gamma(M, \mathcal{O}^{1,0}(X^*)) & \xleftarrow{d} & \Gamma(M, \mathcal{O}(X^*)) & \longleftarrow & \cdots . \end{array}$$

In this diagram, K and L are defined as the kernel and cokernel of the respective mappings d. To show that K and L are dual vector spaces, which is the desired result, it is clearly sufficient to show that this diagram is commutative, in the sense that

$$(F, dG) = (dF, G)$$

for all $F \in H^1(M, \mathcal{O}(X))$ and $G \in \Gamma(M, \mathcal{O}(X^*))$; here, $(\cdot,\cdot)$ denotes the inner product expressing Serre duality. The result is immediate, upon recalling the explicit form of the Serre duality. Letting $\mathfrak{U} = \{U_\alpha\}$ be a Leray covering of M for coherent analytic sheaves, an element $F \in H^1(M, \mathcal{O}(X))$ has a representative cocycle $(F_{\alpha\beta}) \in Z^1(\mathfrak{U}, \mathcal{O}(X))$; and there are C^∞ functions F_α forming a cochain $(F_\alpha) \in C^0(\mathfrak{U}, \mathcal{E}(X))$ such that $(F_{\alpha\beta}) = \delta(F_\alpha)$, that is, such that $F_{\alpha\beta} = F_\beta - \chi_{\alpha\beta}^{-1} F_\alpha$ in $U_\alpha \cap U_\beta$. Then for any section $\psi = (\psi_\alpha) \in \Gamma(M, \mathcal{O}^{1,0}(X^*))$ the dual pairing is defined by

$$(F, \psi) = \int_M {}^t\psi_\alpha \wedge \bar{\partial} F_\alpha .$$

Similarly, an element $\varphi \in H^1(M, \mathcal{O}^{1,0}(X))$ has a representative cocycle $(\varphi_{\alpha\beta}) \in Z^1(\mathfrak{U}, \mathcal{O}^{1,0}(X))$; and there is a cochain $(\varphi_\alpha) \in C^0(\mathfrak{U}, \mathcal{E}^{1,0}(X))$ such that $\varphi_{\alpha\beta} = \varphi_\beta - \chi_{\alpha\beta}^{-1}\varphi_\alpha$ in $U_\alpha \cap U_\beta$.

Then for any section $G = (G_\alpha) \in \Gamma(M, \mathcal{O}(X^*))$, the dual pairing is defined by

$$(\varphi, G) = \int_M {}^tG_\alpha \wedge \bar\partial\varphi_\alpha .$$

Now if $F \in H^1(M, \mathcal{O}(X))$ and $G \in \Gamma(M, \mathcal{O}(X^*))$, select a zero-cochain $(F_\alpha) \in C^0(\mathcal{U}, \mathcal{E}(X))$ with coboundary as the cocycle $(F_{\alpha\beta}) \in Z^1(\mathcal{U}, \mathcal{O}(X))$ representing F ; clearly the zero-cochain $(dF_\alpha) \in C^0(\mathcal{U}, \mathcal{E}^{1,0}(X))$ has as its coboundary the cocycle $(dF_{\alpha\beta}) \in Z^1(\mathcal{U}, \mathcal{O}^{1,0}(X))$ representing dF . Note that ${}^tG_\alpha\bar\partial F_\alpha = {}^tG_\beta\bar\partial F_\beta$ in $U_\alpha \cap U_\beta$, so this is a global differential form ${}^tG_\alpha\bar\partial F_\alpha \in \Gamma(M, \mathcal{E}^{0,1})$. Thus by Stokes' theorem,

$$(F, dG) = \int_M {}^t dG_\alpha \wedge \bar\partial F_\alpha = \int_M (d({}^tG_\alpha\bar\partial F_\alpha) - {}^tG_\alpha d\bar\partial F_\alpha)$$

$$= \int_M {}^tG_\alpha\bar\partial(dF_\alpha) = (dF, G) ,$$

concluding the proof.

Theorem 22. If M is a compact Riemann surface and $X \in H^1(M, GL(n, \mathbf{C}))$ is a flat complex vector bundle over M , there is an exact sequence of complex vector spaces of the form

$$(3) \quad 0 \to \frac{\Gamma(M, \mathcal{O}^{1,0}(X))}{d\Gamma(M, \mathcal{O}(X))} \xrightarrow{\ \delta\ } H^1(M, \mathcal{F}(X)) \xrightarrow{\ \rho\ } \left[\frac{\Gamma(M, \mathcal{O}^{1,0}(X^*))}{d\Gamma(M, \mathcal{O}(X^*))}\right]^* \to 0 .$$

Proof. The exact sequence of vector spaces (2) can be re-written as an exact sequence

$$0 \longrightarrow \Gamma(M, \mathcal{O}^{1,0}(X))/d\Gamma(M, \mathcal{O}(X)) \longrightarrow H^1(M, \mathcal{F}(X)) \longrightarrow K \longrightarrow 0 ,$$

where K is the kernel of the mapping $d\colon H^1(M, \mathcal{O}(X)) \longrightarrow H^1(M, \mathcal{O}^{1,0}(X))$;

and since that kernel K is described as in Lemma 23, the result is demonstrated.

Remarks. The cohomology group $H^1(M, \mathcal{F}(X))$ is in a sense a purely geometrical entity; the complex structure of the Riemann surface M enters only in the form of the exact sequence (3). Clearly the main problem is that of describing explicitly the exact sequence (3), or just that of describing the image of the mapping δ. When X is the identity bundle of rank 1, then $H^1(M, \mathcal{F}(X)) = H^1(M, \mathbf{C})$ is the ordinary cohomology of the surface M; the image of δ consists of those cohomology classes which are represented by the abelian differentials on the surface M, and is described by the period matrix of the surface. For an arbitrary flat vector bundle X, the differential forms $\theta \in \Gamma(M, \mathcal{O}^{1,0}(X))$ will be called (generalized) Prym differentials on the surface M. (Classically, the Prym differentials are such differential forms when X is a unitary flat line bundle on the surface M; see for instance H. Weyl, Die Idee der Riemannschen Fläche.) The mapping δ in (3) is just that which associates to any Prym differential θ its period class, as defined in §7(c); for the Prym differentials are closed differential forms of degree 1 on the surface. The homomorphism ρ in (3) associates to each cohomology class $A \in H^1(M, \mathcal{F}(X))$ a linear functional

$$\rho(A)\colon \Gamma(M, \mathcal{O}^{1,0}(X^*)) \longrightarrow \mathbf{C};$$

and the cohomology class A is the period class of a Prym differential if and only if $\rho(A)$ is the zero mapping. The period classes of the Prym differentials will also be called the holomorphic cohomology classes in $H^1(M, \mathcal{F}(X))$.

In order firmly to fix the conventions, the mappings δ and ρ will be described explicitly as follows. Select an open covering $\mathfrak{U} = \{U_\alpha\}$, which is a Leray covering for coherent analytic sheaves and for flat sheaves. A Prym differential $\theta \in \Gamma(M, \mathcal{O}^{1,0}(X))$ is described by n-tuples θ_α of abelian differential forms in the various open sets U_α , such that $\theta_\alpha = X_{\alpha\beta}\theta_\beta$ in $U_\alpha \cap U_\beta$. In each U_α select a holomorphic function F_α such that $\theta_\alpha = dF_\alpha$; the constants

(4) $$A_{\alpha\beta} = F_\beta - X_{\alpha\beta}^{-1}F_\alpha \quad \text{in} \quad U_\alpha \cap U_\beta$$

form a one-cocycle $(A_{\alpha\beta}) \in Z^1(\mathfrak{U}, \mathcal{F}(X))$, which represents the period class $A = \delta(\theta)$ of the Prym differential. Now an arbitrary cohomology class $A \in H^1(M, \mathcal{F}(X))$ represented by a cocycle $(A_{\alpha\beta}) \in Z^1(\mathfrak{U}, \mathcal{F}(X))$ can be written in the form (4) for some C^∞ functions F_α ; and then $\bar{\partial}F_\alpha \in \Gamma(M, \mathcal{E}^{0,1}(X))$. For any Prym differential $\varphi \in \Gamma(M, \mathcal{O}^{1,0}(X^*))$ of the dual flat vector bundle X^* , the mapping $\rho(A)$ associates to φ the value

(5) $$\rho(A)\cdot\varphi = \int_M {}^t\varphi_\alpha \wedge \bar{\partial}F_\alpha \, .$$

If A is a holomorphic cohomology class, the function F_α can be taken to be holomorphic, so $\bar{\partial}F_\alpha = 0$ and $\rho(A)\cdot\varphi = 0$ for all φ ; conversely, by the exactness of the sequence (3), if $\rho(A)\cdot\varphi = 0$ for all $\varphi \in \Gamma(M, \mathcal{O}^{1,0}(X^*))$, then the cocycle $(A_{\alpha\beta})$ can be represented in the form (4) where F_α is holomorphic.

The explicit mappings in the exact sequence (3) can also be described in the following slightly different form. In additon to (3), consider the corresponding exact sequence for the dual vector bundle X^*; and write the two sequences out in the following form.

$$
\begin{array}{ccccccccc}
0 & \longrightarrow & \dfrac{\Gamma(M,\mathcal{O}^{1,0}(X))}{d\Gamma(M,\mathcal{O}(X))} & \xrightarrow{\delta} & H^1(M,\mathcal{J}(X)) & \xrightarrow{\rho} & \left[\dfrac{\Gamma(M,\mathcal{O}^{1,0}(X^*))}{d\Gamma(M,\mathcal{O}(X^*))}\right]^* & \longrightarrow & 0 \\
(6) & & \updownarrow \alpha & & \updownarrow \beta & & \updownarrow \gamma & & \\
0 & \longleftarrow & \left[\dfrac{\Gamma(M,\mathcal{O}^{1,0}(X))}{d\Gamma(M,\mathcal{O}(X))}\right]^* & \xleftarrow{\rho} & H^1(M,\mathcal{J}(X^*)) & \xleftarrow{\delta} & \dfrac{\Gamma(M,\mathcal{O}^{1,0}(X^*))}{d\Gamma(M,\mathcal{O}(X^*))} & \longleftarrow & 0
\end{array}
$$

There are the dual pairings α and γ in (6), which are the obvious ones. In addition, it is clear that the operation of complex conjugation is an isomorphism (or more precisely, a conjugate-linear isomorphism) $H^1(M, \mathcal{J}(X^*)) \cong H^1(M, \mathcal{J}(\overline{X}^*))$; combining this with the Hermitian dual pairing

$$H^1(M, \mathcal{J}(X)) \otimes H^1(M, \mathcal{J}(\overline{X}^*)) \longrightarrow \mathbb{C}$$

of Theorem 20 establishes a dual pairing β in (6). To be explicit, if $A \in H^1(M, \mathcal{J}(X))$ and $B \in H^1(M, \mathcal{J}(X^*))$, the dual pairing β is defined by

$$(7) \qquad [A,B] = \langle A,\overline{B} \rangle ,$$

where $\overline{B} \in H^1(M, \mathcal{J}(\overline{X}^*))$ is the complex conjugate cohomology class to B, and $\langle\cdot,\cdot\rangle$ is defined as in equation (11) of §7.

<u>Lemma 24.</u> Over a compact Riemann surface, (6) is a commutative diagram, in the sense that

$$\rho(B)\cdot\varphi = [B,\delta\varphi] \quad \text{and} \quad \rho(A)\cdot\psi = [A,\delta\psi]$$

whenever $\varphi \in \Gamma(M, \mathcal{O}^{1,0}(X))$, $\psi \in \Gamma(M, \mathcal{O}^{1,0}(X^*))$, $A \in H^1(M, \mathcal{F}(X))$, and $B \in H^1(M, \mathcal{F}(X^*))$.

Proof. Since the diagram (6) is symmetric, it suffices to prove commutativity only on one side. Consider then a cohomology class $A \in H^1(M, \mathcal{F}(X))$, represented by a cocycle $(A_{\alpha\beta}) \in Z^1(\mathfrak{U}, \mathcal{F}(X))$, and a Prym differential $\psi = (\psi_\alpha) \in \Gamma(M, \mathcal{O}^{1,0}(X^*))$; select C^∞ functions F_α satisfying $A_{\alpha\beta} = F_\beta - X_{\alpha\beta}^{-1}F_\alpha$ in $U_\alpha \cap U_\beta$, and let $B = \delta\psi \in H^1(M, \mathcal{F}(X^*))$. Then, from the discussion on page 136 and from (5), it follows that

$$[A,B] = < A,\overline{B} > = \int_M {}^t\psi_\alpha \wedge dF_\alpha$$
$$= \int_M {}^t\psi_\alpha \wedge \overline{\partial}F_\alpha = \rho(A)\cdot\psi ,$$

as desired.

Note now that the condition that a cohomology class $A \in H^1(M, \mathcal{F}(X))$ be analytic, that is, that it be the period class of a Prym differential in $\Gamma(M, \mathcal{O}^{1,0}(X))$, is just that $0 = \rho(A)\cdot\psi = [A,\delta\psi]$ for all the Prym differentials $\psi \in \Gamma(M, \mathcal{O}^{1,0}(X^*))$. The dual pairing $[A,\delta\psi]$ is given explicitly, in terms of the flat cocycles, by Theorem 21.

(b) The exact sequence (3), for the special case that X is the identity bundle of rank 1, plays an important role in the treatment of the Picard-Jacobi variety attached to a compact Riemann surface. It is perhaps of some interest to digress briefly, to see the extent to which some of the standard observations about the special case of that sequence carry over to the general case.

When X is the identity bundle, $H^1(M, \mathcal{F}(X)) = H^1(M, \mathbb{C})$ is an even-dimensional complex vector space; and the exact sequence (3) is an even splitting of the vector space $H^1(M, \mathcal{F}(X))$, in the sense that

$$(8) \qquad \dim \frac{\Gamma(M, \mathcal{O}^{1,0}(X))}{d\Gamma(M, \mathcal{O}(X))} = \dim \frac{\Gamma(M, \mathcal{O}^{1,0}(X^*))}{d\Gamma(M, \mathcal{O}(X^*))} .$$

In general, the space $H^1(M, \mathcal{F}(X))$ need not be even-dimensional; the condition that this be so is of course just that $p+q$ be even, with the notation as in the Corollary to Theorem 19 (page 133). Even when $H^1(M, \mathcal{F}(X))$ is even-dimensional, the exact sequence (3) need not be an even splitting. For some special cases, however, the even splitting does occur; these results are summarized in the next theorem. Note firstly, however, that the condition (8) can be simplified somewhat. On the one hand, recalling the exact sequence (2), it follows that $\dim d\Gamma(M, \mathcal{O}(X)) = \dim \Gamma(M, \mathcal{O}(X)) - \dim \Gamma(M, \mathcal{F}(X))$; and on the other hand, from the Riemann-Roch theorem it follows that $\dim \Gamma(M, \mathcal{O}(X)) - \dim \Gamma(M, \mathcal{O}^{1,0}(X^*)) = n(1-g)$, where n is the rank of the bundle X and g is the genus of the surface M . Thus the condition (8) for an even splitting can be rewritten as the condition

$$(9) \qquad 2 \dim \Gamma(M, \mathcal{O}(X)) - \dim \Gamma(M, \mathcal{F}(X)) = 2 \dim \Gamma(M, \mathcal{O}(X^*)) - \dim \Gamma(M, \mathcal{F}(X^*)) .$$

Theorem 23. Let M be a compact Riemann surface, and $X \in H^1(M, GL(n, \mathbf{C}))$ be a flat vector bundle of rank n over that surface.

(i) If $n = 1$, the exact sequence (3) is always an even splitting.

(ii) If $n = 2$ and X is a stable vector bundle, the exact sequence (3) is always an even splitting.

(iii) If $n = 2$ and X is an analytically indecomposable unstable vector bundle, let ξ be the unique complex line bundle such that $c(\xi) = \operatorname{div} X$ and $\xi \subset X$. Then the exact sequence (3) is an even splitting if and only if

$$2 \dim \Gamma(M, \mathcal{O}(\xi)) - \dim \Gamma(M, \vartheta(\xi)) = \\ = 2 \dim \Gamma(M, \mathcal{O}(\xi \otimes \det X^*)) - \dim \Gamma(M, \vartheta(\xi \otimes \det X^*)) .$$

(iv) If X is a unitary flat vector bundle of any rank, the exact sequence (3) is always an even splitting.

Proof. (i) When $n = 1$, it follows from the Riemann-Roch Theorem for complex line bundles that

$$\dim \Gamma(M, \mathcal{O}(X)) = \begin{cases} 1 & \text{if } X = 1 \text{ in } H^1(M, \mathcal{O}^*) \\ 0 & \text{otherwise} \end{cases} ;$$

and it follows from Theorem 18 that

$$\dim \Gamma(M, \vartheta(X)) = \begin{cases} 1 & \text{if } X = 1 \text{ in } H^1(M, \mathbf{C}^*) \\ 0 & \text{otherwise.} \end{cases}$$

Since in either cohomology group $X = 1$ if and only if $X^* = 1$, condition (9) is verified and the desired result proved in this case.

(ii) For a flat vector bundle X of course $c(\det X) = 0$; thus X is stable precisely when $c(\xi) < 0$ for any complex analytic line bundle such that $\xi \subset X$. Recalling (10) in §5 for instance, it follows that $\Gamma(M, \mathcal{O}(X)) = 0$ whenever X is stable; and necessarily $\Gamma(M, \mathcal{J}(X)) = 0$ as well. In view of condition (9), it is then sufficient to observe that X <u>is stable precisely when</u> X^* <u>is stable.</u> To see this, consider any line bundle $\xi \subset X$; the analytic line bundle ξ and the analytic vector bundle X can then be represented by cocycles $(\xi_{\alpha\beta})$ and $(X_{\alpha\beta})$ in terms of a suitable covering $\mathfrak{U} = \{U_\alpha\}$ of M, such that

$$X_{\alpha\beta} = \begin{pmatrix} \xi_{\alpha\beta} & \lambda_{\alpha\beta} \\ 0 & \eta_{\alpha\beta} \end{pmatrix}, \tag{10}$$

for some line bundle $\eta = (\eta_{\alpha\beta})$. The dual bundle X^* is then represented by the cocycle $X^*_{\alpha\beta}$, where

$$X^*_{\alpha\beta} = \begin{pmatrix} \xi^{-1}_{\alpha\beta} & 0 \\ \lambda'_{\alpha\beta} & \eta^{-1}_{\alpha\beta} \end{pmatrix};$$

upon interchanging rows and columns, the same bundle is represented by the cocycle

$$X'_{\alpha\beta} = \begin{pmatrix} \eta^{-1}_{\alpha\beta} & \lambda'_{\alpha\beta} \\ 0 & \xi^{-1}_{\alpha\beta} \end{pmatrix}. \tag{11}$$

It is thus apparent that $\xi \subset X$ precisely when $\eta^{-1} = \xi(\det X)^{-1} \subset X^*$; and since $c(\eta^{-1}) = c(\xi)$, it follows immediately that X is stable precisely when X^* is stable, which proves the desired result in this case.

(iii) The fact that there is a unique complex analytic line bundle ξ such that $c(\xi) = \operatorname{div} X \geqq 0$ and $\xi \subset X$ follows from Lemma 15. In terms of a suitable covering $\mathfrak{U} = \{U_\alpha\}$ of M, the analytic bundles ξ and χ can be defined by cocycles $(\xi_{\alpha\beta})$ and $(\chi_{\alpha\beta})$, where $\chi_{\alpha\beta}$ is as in (10); the dual bundle χ^* can then be defined by a cocycle $(\chi'_{\alpha\beta})$ as in (11). Note that χ^* is unstable as well; and $\eta^{-1} = \xi \det \chi^*$ is the unique complex line bundle such that $c(\eta^{-1}) = \operatorname{div} \chi^* \geqq 0$ and $\eta^{-1} \subset \chi^*$. Now consider an analytic section

$$F_\alpha = \begin{pmatrix} f_\alpha \\ g_\alpha \end{pmatrix} \in \Gamma(M, \mathcal{O}(\chi)) .$$

Since $g_\alpha \in \Gamma(M, \mathcal{O}(\eta))$ and $c(\eta) = -c(\xi) \leqq 0$, it follows that $g_\alpha = 0$ unless $\eta = 1$ in $H^1(M, \mathcal{O}^*)$; but if $\eta = 1$ and $g_\alpha \neq 0$, it follows as in the proof of part (ii) of Lemma 15 that the bundle χ is analytically decomposable, which is impossible by assumption. Thus necessarily $g_\alpha = 0$, and so $\Gamma(M, \mathcal{O}(\chi)) \cong \Gamma(M, \mathcal{O}(\xi))$; and then $\Gamma(M, \mathcal{F}(\chi)) \cong \Gamma(M, \mathcal{F}(\xi))$ as well. Since $\eta^{-1} = \xi \otimes \det \chi^*$ plays the same role for the dual bundle χ^*, the desired conclusion in this case follows immediately from condition (9).

(iv) If χ is a unitary flat vector bundle, then the complex analytic sections of χ are necessarily all constants; (see the following Lemma 25). That is to say, $\Gamma(M, \mathcal{O}(\chi)) = \Gamma(M, \mathcal{F}(\chi))$; therefore the condition (9) that there be an even splitting in the exact sequence (3) reduces to the condition that $\dim \Gamma(M, \mathcal{F}(\chi)) = \dim \Gamma(M, \mathcal{F}(\chi^*))$. Now from Theorem 18 it follows that $r = \dim \Gamma(M, \mathcal{F}(\chi))$ is the largest integer such that the identity representation of degree r be contained in the unitary representation $\hat{\chi}$ of the group $\pi_1(M)$; it is familiar, however, that reducible unitary representations are fully decomposable, so that $r = \dim \Gamma(M, \mathcal{F}^{\cdot}(X^*))$ as well. The proof of the theorem is then completed, except for the proof of the following result.

Lemma 25. If X is a unitary flat vector bundle over a compact Riemann surface M, the complex analytic sections of X are necessarily all constants; that is to say, $\Gamma(M, \mathcal{O}(X)) = \Gamma(M, \mathcal{F}(X))$.

Proof. Consider a section $(F_\alpha) \in \Gamma(M, \mathcal{O}(X))$ expressed in terms of an open covering $\mathcal{U} = \{U_\alpha\}$ of the surface M; the elements F_α will be viewed as column vectors of holomorphic functions on U_α of length n, where n is the rank of X. Introduce the norm $\|F_\alpha\| = {}^t\overline{F}_\alpha \cdot F_\alpha$, which is a real-valued function in U_α; since the matrices $X_{\alpha\beta}$ of a cocycle defining the bundle X are unitary, it follows that $\|F_\alpha\| = \|F_\beta\|$ in $U_\alpha \cap U_\beta$, hence that $\|F_\alpha\|$ is a well-defined, real-valued function on the entire surface M. Since M is compact, this function will attain its maximum at some point $p_o \in U_{\alpha_o} \subset M$; upon multiplying all the vectors F_α by the same unitary constant matrix, it can be assumed that

$$F_{\alpha_o}(p_o) = \begin{pmatrix} f_{1\alpha_o}(p_o) \\ 0 \\ \vdots \\ 0 \end{pmatrix}.$$

Now the function $f_{1\alpha_o}(p)$ is holomorphic in the open neighborhood

U_{α_o} of p_o, and

$$|f_{1\alpha_o}(p)| \leqq \|F_{\alpha_o}(p)\| \leqq \|F_{\alpha_o}(p_o)\| = |f_{1\alpha_o}(p_o)| \quad \text{for all} \quad p \in U_{\alpha_o};$$

so by the maximum modulus theorem, $f_{1\alpha_o}(p)$ is constant in U_{α_o}. Furthermore, since $f_{1\alpha_o}(p) \equiv f_{1\alpha_o}(p_o)$ and $\|F_{\alpha_o}(p)\| \leqq \|F_{\alpha_o}(p_o)\|$ for all $p \in U_{\alpha_o}$, necessarily $f_{2\alpha_o}(p) \equiv \dots \equiv f_{n\alpha_o}(p) \equiv 0$ in U_{α_o} as well; thus F_{α_o} is constant in U_{α_o}, hence is constant on the entire (connected) Riemann surface M, as desired.

The observation that stable and unitary bundles behave similarly will appear in its true light later, when the relations between these two classes of bundles are discussed in more detail. Upon considering case (iii) of Theorem 23 more closely, it is easy to construct examples of such bundles for which the exact sequence (3) is not an even splitting. Let ξ, η be any two complex analytic line bundles over the surface M such that $0 < c(\xi) = c(\eta^{-1})$, that $\dim \Gamma(M, \mathcal{O}(\xi)) \neq \dim \Gamma(M, \mathcal{O}(\eta^{-1}))$, and that there is an indecomposable complex analytic vector bundle X which is an extension of ξ by η. (That there exist such line bundles on surfaces of genus $g \geqq 4$ is quite trivial. Select any two line bundles ξ, η such that $1 \leqq c(\xi) = c(\eta^{-1})$ and that $\dim \Gamma(M, \mathcal{O}(\xi)) \neq \dim \Gamma(M, \mathcal{O}(\eta^{-1}))$; for instance, select for ξ a point bundle $\xi = \zeta_p$, and select for η^{-1} a bundle of Chern class $c(\eta^{-1}) = 1$ which is not a point bundle. The set of all extensions of ξ by η is in one-to-one correspondence with the cohomology group $H^1(M, \mathcal{O}(\xi\eta^{-1}))$, by Theorem 13; so to guarantee the existence of a non-trivial extension, it suffices merely to ensure that that cohomology group be non-trivial. By Serre's duality theorem, $H^1(M, \mathcal{O}(\xi\eta^{-1})) \cong \Gamma(M, \mathcal{O}(\kappa\xi^{-1}\eta))$; and since $c(\kappa\xi^{-1}\eta) = 2g-2-2c(\xi)$, it follows from the Riemann-Roch theorem for line bundles (see page 112 in last year's lectures) that $\Gamma(M, \mathcal{O}(\kappa\xi^{-1}\eta)) \neq 0$ whenever $c(\kappa\xi^{-1}\eta) = 2g-2-2c(\xi) \geqq g$, hence whenever $2c(\xi) \leqq g-2$.) Since $c(\xi) = c(\eta^{-1}) > 0$, it is clear that $\Gamma(M, \mathcal{J}(\xi)) = \Gamma(M, \mathcal{J}(\eta^{-1})) = 0$; so in view of the criterion of part (iii) of Theorem 23, the exact sequence (3) is evidently not an even splitting for this vector bundle X . On the other hand, though, there are analytically indecomposable unstable flat vector bundles X such that the exact sequence (3) is an even splitting. For instance, let X be any such bundle for which $\text{div}\, X = g-1$, where $g > 1$ is the genus of the surface M . Let ξ be that line bundle for which $c(\xi) = \text{div}\, X = g-1$ and $\xi \subset X$; and let $\eta = \xi^{-1} \det X$, so that X is a non-trivial extension of ξ by η . By Theorem 13, there exists such an extension if and only if $H^1(M, \mathcal{O}(\xi\eta^{-1})) \cong \Gamma(M, \mathcal{O}(\kappa\xi^{-1}\eta)) \neq 0$, hence if and only $\xi = \kappa\eta$, since $c(\kappa\xi^{-1}\eta) = 2g-2-2c(\xi) = 0$. Now clearly $\Gamma(M, \mathcal{J}(\xi)) = \Gamma(M, \mathcal{J}(\eta)) = 0$; and from the Riemann-Roch theorem, $\dim \Gamma(M, \mathcal{O}(\xi)) = \dim \Gamma(M, \mathcal{O}(\kappa\xi^{-1})) + c(\xi) + 1-g = \dim \Gamma(M, \mathcal{O}(\eta^{-1}))$. Therefore, by the criterion of part (iii) of Theorem 23, the exact sequence (3) is an even splitting for this vector bundle X . This latter particular case is of some interest in uniformization questions, as noted in last year's lectures.

Again, when X is the identity bundle, the exact sequence

(3) is an even splitting, and moreover, the cohomology group $H^1(M, \mathcal{F}(X))$ is the direct sum of the image of δ and of its complex conjugate. In general, the Prym differentials in $\Gamma(M, \mathcal{O}^{1,0}(\overline{X}))$ have periods in $H^1(M, \mathcal{F}(\overline{X}))$; and complex conjugation establishes a conjugate-linear isomorphism $H^1(M, \mathcal{F}(\overline{X})) \cong H^1(M, \mathcal{F}(X))$. Combining these two mappings leads to the isomorphism (into)

$$\overline{\delta}: \frac{\Gamma(M, \mathcal{O}^{1,0}(\overline{X}))}{d\Gamma(M, \mathcal{O}(\overline{X}))} \longrightarrow H^1(M, \mathcal{F}(X)) ;$$

and one can ask whether there is an isomorphism

$$H^1(M, \mathcal{F}(X)) \cong (\text{im } \delta) \oplus (\text{im } \overline{\delta}) , \tag{12}$$

where im δ denotes the image of the homomorphism δ .

Theorem 24. If M is a compact Riemann surface and $X \in H^1(M, GL(n,\mathbf{C}))$ is a unitary flat vector bundle over M , then there is an isomorphism of the form (12).

Proof. It follows immediately from Theorem 23(iv) that $\dim (\text{im } \delta) = \dim(\text{im } \overline{\delta}) = \frac{1}{2} \dim H^1(M, \mathcal{F}(X))$; so it is merely necessary to show that $(\text{im } \delta) \cap (\text{im } \overline{\delta}) = 0$. Suppose contrariwise that there is a non-trivial cohomology class $A \in (\text{im } \delta) \cap (\text{im } \overline{\delta}) \subset H^1(M, \mathcal{F}(X))$. In terms of a suitable open covering $\mathcal{U} = \{U_\alpha\}$ of the surface M that cohomology class is represented by a cocycle $(A_{\alpha\beta}) \in Z^1(\mathcal{U}, \mathcal{F}(X))$ such that

$$A_{\alpha\beta} = F_\beta - X_{\alpha\beta}^{-1} F_\alpha = \overline{G}_\beta - X_{\alpha\beta}^{-1} \overline{G}_\alpha \quad \text{in} \quad U_\alpha \cap U_\beta ,$$

for some cochains $(F_\alpha) \in C^0(\mathcal{U}, \mathcal{O}(X))$, $(G_\alpha) \in C^0(\mathcal{U}, \mathcal{O}(\overline{X}))$;

thus $A = \delta(dF_\alpha) = \overline{\delta}(dG_\alpha)$. Now the functions $H_\alpha = F_\alpha - \overline{G}_\alpha$ are harmonic in U_α , and $H_\alpha = \chi_{\alpha\beta}H_\beta$ in $U_\alpha \cap U_\beta$; harmonic functions also satisfy the maximum modulus principle though, so as in the proof of Lemma 25 it follows that H_α is constant. Since $F_\alpha = \overline{G}_\alpha + H_\alpha$ and both F_α and G_α are holomorphic, it then follows that both F_α and G_α are also constants; but then the cohomology class A is trivial, which is a contradiction. That serves to complete the proof.

Remark. If χ is a real unitary (orthogonal) bundle, then $\overline{\chi} = \chi$, and it follows that $(\text{im } \overline{\delta}) = (\overline{\text{im } \delta})$; this is true in particular when χ is the identity bundle.

Finally, there is the question of the construction of generalizations of the Picard variety of a Riemann surface, involving the cohomology groups $H^1(M, \mathcal{F}(\chi))$. In many cases of arithmetic interest, where the representation $\hat{\chi}$ is essentially rational, it is possible to define a lattice subgroup of $H^1(M, \mathcal{F}(\chi))$ and parallel the construction of the Picard variety, as given in last year's lectures. [See the paper by Goro Shimura, "Sur les intégrales attachèes aux formes automorphes," Jour. Math. Soc. Japan 11(1959), 291-311.] This method fails in the general case; we shall return to this question later, from another point of view.

(c) The study of the analytic properties of flat sheaves can be approached in a slightly different way, by considering instead of the exact sequence (1) the following exact sequence of sheaves:

$$(13) \qquad 0 \longrightarrow \mathcal{F}(\chi) \longrightarrow \mathcal{M}(\chi) \xrightarrow{d} \mathcal{M}^{1,0}(\chi) ,$$

where $\mathcal{M}$ denotes the sheaf of germs of meromorphic functions on the Riemann surface M. The mapping d in (13) is not onto the entire sheaf $\mathcal{M}^{1,0}(X)$ of germs of meromorphic differential forms which are sections of the vector bundle X; as is familiar, the image $d\mathcal{M}(X) \subset \mathcal{M}^{1,0}(X)$ consists of those germs which have zero residues. The exact cohomology sequence associated to (13) has the form

$$(14) \qquad 0 \longrightarrow \Gamma(M, \mathcal{F}(X)) \longrightarrow \Gamma(M, \mathcal{M}(X)) \xrightarrow{d} \Gamma(M, d\mathcal{M}(X)) \longrightarrow$$

$$\longrightarrow H^1(M, \mathcal{F}(X)) \longrightarrow 0 ,$$

since $H^1(M, \mathcal{M}(X)) = 0$ as noted earlier. Therefore the flat cohomology has the analytic representation

$$(15) \qquad H^1(M, \mathcal{F}(X)) \cong \frac{\Gamma(M, d\mathcal{M}(X))}{d\Gamma(M, \mathcal{M}(X))} ;$$

indeed, this holds for an arbitrary Riemann surface M, although we shall consider here only compact Riemann surfaces.

For any flat vector bundle X, the sections $\theta \in \Gamma(M, d\mathcal{M}(X))$ will be called meromorphic Prym differentials on the surface M. It should be emphasized that the meromorphic Prym differentials are those meromorphic differential forms $\theta \in \Gamma(M, \mathcal{M}^{1,0}(X))$ which have zero residues at each point of the surface; in the classical terminology, these would be known as differential forms of the second kind. With this in mind, these meromorphic Prym differentials θ have well-defined period classes $\delta\theta \in H^1(M, \mathcal{F}(X))$, as in §7(c); the period mapping $\delta\colon \Gamma(M, d\mathcal{M}(X)) \longrightarrow H^1(M, \mathcal{F}(X))$ is precisely the mapping arising in the exact cohomology sequence (14). Thus the analytic representation (14) can be interpreted as the assertion that each cohomology class $A \in H^1(M, \mathcal{F}(X))$ is the period class of some meromorphic Prym differential on the surface M; and that two meromorphic Prym differentials have the same periods precisely when their difference is the exterior derivative of a meromorphic section $f \in \Gamma(M, \mathcal{M}(X))$. The distinguished subspace of $H^1(M, \mathcal{F}(X))$ of greatest interest is the space of those cohomology classes which are the periods of holomorphic Prym differentials.

There is a slight variation of this analytic representation of the flat cohomology based on the observation, made earlier during these lectures, that all vector bundles are meromorphically trivial. Let $\mathcal{U} = \{U_\alpha\}$ be an open covering of the Riemann surface M which is a Leray covering both for flat sheaves and for analytic sheaves; and select a representative cocycle $(X_{\alpha\beta}) \in Z^1(\mathcal{U}, GL(n, \mathbf{C}))$ for the flat vector bundle X. There are elements $P_\alpha \in GL(n, \mathcal{M}_{U_\alpha}) = \Gamma(U_\alpha, \mathcal{GL}(n, \mathcal{M}))$ such that $P_\alpha(z) = X_{\alpha\beta} \cdot P_\beta(z)$ whenever $z \in U_\alpha \cap U_\beta$; and in terms of these matrices the isomorphism $P: \mathcal{M}(X) \longrightarrow \mathcal{M}^n$ is given by

$$P(F_\alpha) = P_\alpha^{-1} F_\alpha \quad \text{for} \quad F_\alpha \in \mathcal{M}(X) .$$

The same mapping of course yields the isomorphism $P: \mathcal{M}^{1,0}(X) \longrightarrow (\mathcal{M}^{1,0})^n$. Applying these isomorphisms to the exact sequence (13), there results the commutative diagram of exact sequences

$$
\begin{array}{ccccccc}
0 \longrightarrow & \mathcal{F}(X) & \longrightarrow & \mathcal{M}(X) & \xrightarrow{d} & \mathcal{M}^{1,0}(X) & \\
 & \text{id.}\downarrow\cong & & P\downarrow\cong & & P\downarrow\cong & \\
0 \longrightarrow & \mathcal{F}(X) & \xrightarrow{i_P} & \mathcal{M} & \xrightarrow{d_P} & \mathcal{M}^{1,0} & ;
\end{array} \tag{16}
$$

note that i_P is just the mapping P itself, while d_P is the mapping given by

$$
\begin{aligned}
d_P(F_\alpha) &= P\cdot d\cdot P^{-1}(F_\alpha) = P_\alpha^{-1}d(P_\alpha F_\alpha) \\
&= dF_\alpha + P_\alpha^{-1}dP_\alpha\cdot F_\alpha ,
\end{aligned} \tag{17}
$$

where $F_\alpha \in \mathcal{M}$ and d is the ordinary exterior derivative. The exact cohomology sequence associated to the second line in (16) has the form

$$
\begin{aligned}
0 \longrightarrow \Gamma(M,\mathcal{F}(X)) \xrightarrow{i_P} \Gamma(M,\mathcal{M}) \xrightarrow{d_P} \Gamma(M,d_P\mathcal{M}) \xrightarrow{\delta} \\
\longrightarrow H^1(M,\mathcal{F}(X)) \longrightarrow 0 \quad ,
\end{aligned}
$$

since again $H^1(M,\mathcal{M}) = 0$; therefore the flat cohomology has the analytic representation

$$
H^1(M,\mathcal{F}(X)) = \frac{\Gamma(M,d_P\mathcal{M})}{d_P\Gamma(M,\mathcal{M})} \quad . \tag{18}
$$

The advantage of this representation is that the sections are global meromorphic differential forms on the Riemann surface; the vector bundle appears only in the differential operator d_P .

The dual pairing $H^1(M,\mathcal{F}(X))\otimes H^1(M,\mathcal{F}(X^*)) \longrightarrow \mathbb{C}$ considered on page 162, which assigns to cohomology classes

$A \in H^1(M, \mathcal{F}(X))$ and $B \in H^1(M, \mathcal{F}(X^*))$ the complex number denoted by $[A,B]$, has an interesting form in terms of this analytic representation of the flat cohomology. If the cohomology classes A,B are represented by $\varphi \in \Gamma(M, d\mathcal{M}(X))$ and $\psi \in \Gamma(M, d\mathcal{M}(X^*))$ respectively under the isomorphism (15), recall that in a suitable coordinate neighborhood of any point, φ and ψ are represented by n-tuples of ordinary meromorphic differential forms, viewed as column vectors; moreover these vector-valued meromorphic differential forms can be expressed locally as the exterior derivatives of some vector-valued meromorphic functions, which functions will be denoted by $\int\varphi$ and $\int\psi$.

Theorem 25. On a compact Riemann surface M the dual pairing $H^1(M, \mathcal{F}(X)) \otimes H^1(M, \mathcal{F}(X^*)) \longrightarrow \mathbf{C}$ has the form

$$[A,B] = 2\pi i\, \mathcal{R}[{}^t(\textstyle\int\psi)\cdot\varphi] = -2\pi i\, \mathcal{R}[{}^t\psi\cdot(\textstyle\int\varphi)], \tag{19}$$

where $\varphi \in \Gamma(M, d\mathcal{M}(X))$ and $\psi \in \Gamma(M, d\mathcal{M}(X^*))$ represent the cohomology classes $A \in H^1(M, \mathcal{F}(X))$ and $B \in H^1(M, \mathcal{F}(X^*))$ respectively under the isomorphism (15), and $\mathcal{R}[\cdot]$ denotes the total residue of the differential form in brackets.

Proof. Note that the residues in (19) are independent of the choices of the integrals of the differential forms; for any two integrals differ by a constant, and the differential forms φ, ψ both have zero residues at each point. Now consider cohomology classes $A \in H^1(M, \mathcal{F}(X))$, $B \in H^1(M, \mathcal{F}(X^*))$, and representative cocycles $(A_{\alpha\beta}) \in Z^1(\mathcal{U}, \mathcal{F}(X))$, $(B_{\alpha\beta}) \in Z^1(\mathcal{U}, \mathcal{F}(X^*))$ respectively, in terms of a suitable open covering $\mathcal{U} = \{U_\alpha\}$ of the surface M. The differential forms $\varphi \in \Gamma(M, d\mathcal{M}(X))$, $\psi \in \Gamma(M, d\mathcal{M}(X^*))$ representing A, B respectively under the isomorphism (15) are given by $\varphi_\alpha = dF_\alpha$, $\psi_\alpha = dG_\alpha$ in U_α, when F_α, G_α are meromorphic vector-valued functions in U_α such that $A_{\alpha\beta} = F_\beta - X_{\alpha\beta}^{-1}F_\alpha$, $B_{\alpha\beta} = G_\beta - (X_{\alpha\beta}^*)^{-1}G_\alpha$ in $U_\alpha \cap U_\beta$. The covering $\mathcal{U}$ can be so chosen that each singularity of F_α or G_α has an open neighborhood meeting no other set of the covering $\mathcal{U}$ but U_α; and modifying these functions F_α, G_α in these neighborhoods of their singularities leads in the obvious manner to C^∞ functions F'_α, G'_α in U_α which also satisfy the relations $A_{\alpha\beta} = F'_\beta - X_{\alpha\beta}^{-1}F'_\alpha$, $B_{\alpha\beta} = G'_\beta - (X_{\alpha\beta}^*)^{-1}G'_\alpha$ in $U_\alpha \cap U_\beta$, since $F_\alpha = F'_\alpha$, etc., in $U_\alpha \cap U_\beta$. The differential forms $\varphi'_\alpha = dF'_\alpha \in \Gamma(M, \mathcal{E}^1(X))$, $\psi'_\alpha = dG'_\alpha \in \Gamma(M, \mathcal{E}^1(X^*))$ then represent the cohomology classes A, B respectively, under the deRham isomorphism of §7(c). Recalling pages 137 and 162, the dual pairing of the Theorem is given by

$$[A,B] = \langle A, \overline{B} \rangle = \int_M {}^t\psi'_\alpha \wedge \varphi'_\alpha = \sum_\alpha \int_{U_\alpha} {}^t\psi'_\alpha \wedge \varphi'_\alpha ,$$

since ${}^t\psi'_\alpha \wedge \varphi'_\alpha = {}^t\psi_\alpha \wedge \varphi_\alpha = 0$ in $U_\alpha \cap U_\beta$ for any other set U_β of the covering. Assuming the sets U_α have smooth boundaries, it follows from Stokes' Theorem that

$$\int_{U_\alpha} {}^t\psi'_\alpha \wedge \varphi'_\alpha = \int_{U_\alpha} {}^t(dG'_\alpha) \wedge \varphi'_\alpha = \int_{\partial U_\alpha} {}^tG'_\alpha \cdot \varphi'_\alpha ;$$

but since $G_\alpha = G'_\alpha$ and $\varphi_\alpha = \varphi'_\alpha$ on ∂U_α, it further follows that

$$\int_{\partial U_\alpha} {}^tG'_\alpha \cdot \varphi'_\alpha = \int_{\partial U_\alpha} {}^tG_\alpha \cdot \varphi_\alpha = 2\pi i \, \mathcal{R}_{U_\alpha}[{}^tG_\alpha \cdot \varphi_\alpha] ,$$

where $\mathcal{R}_{U_\alpha}[\cdot]$ denotes the total residue in U_α. Altogether then,

$$[A,B] = \sum_{\alpha} 2\pi i\, \mathcal{R}_{U_\alpha}[{}^tG_\alpha \cdot \varphi_\alpha] = 2\pi i\, \mathcal{R}[{}^tG_\alpha \cdot \varphi_\alpha] ,$$

since there are no residues in the intersections $U_\alpha \cap U_\beta$. Noting that $[A,B] = -[B,A]$, this can be rewritten

$$[A,B] = -2\pi i\, \mathcal{R}[{}^t\psi_\alpha \cdot F_\alpha] ,$$

and this suffices to conclude the proof.

§9. Families of flat vector bundles

(a) On any surface M, the mapping which associates to a flat vector bundle X its characteristic representation $\hat{X}$ establishes a one-to-one correspondence

$$(1) \qquad H^1(M, GL(n,\mathbb{C})) \longrightarrow \mathrm{Hom}(\pi_1(M), GL(n,\mathbb{C}))/GL(n,\mathbb{C}) \; ;$$

and this can be used to describe in a reasonably convenient manner the family of all flat vector bundles over the surface. Moreover, this description provides a natural complex analytic structure associated to the family of flat vector bundles over a compact surface.

Suppose that M is a compact Riemann surface of genus g; as noted earlier, the fundamental group $\pi_1(M)$ can be described as a group with $2g$ generators $\sigma_1,\ldots,\sigma_g,\ \tau_1,\ldots,\tau_g$ and one relation

$$(2) \qquad [\sigma_g,\tau_g] \cdots [\sigma_2,\tau_2][\sigma_1,\tau_1] = 1 \; ,$$

where the commutator is written $[\sigma,\tau] = \sigma\tau\sigma^{-1}\tau^{-1}$, as usual. Correspondingly, an element $\rho \in \mathrm{Hom}(\pi_1(M), GL(n,\mathbb{C}))$ is completely described by the $2g$ matrices

$$\left.\begin{array}{l} \rho(\sigma_j) = S_j \in GL(n,\mathbb{C}) \\ \rho(\tau_j) = T_j \in GL(n,\mathbb{C}) \end{array}\right\} \; j = 1,\ldots,g \; ;$$

and these can be arbitrary matrices, subject only to the restriction that

$$[S_g,T_g] \cdots [S_2,T_2][S_1,T_1] = I \; .$$

Thus the set $\text{Hom}(\pi_1(M), GL(n,\mathbb{C}))$ can be identified with a certain subset of the product space $GL(n,\mathbb{C})^{2g}$. Recall that the group $GL(n,\mathbb{C})$ has the natural structure of a complex analytic manifold of complex dimension n^2; (see for instance C. Chevalley, Theory of Lie Groups I, (Princeton University Press, 1946)). Explicitly, the mapping $Z \longrightarrow A\cdot\exp Z$ takes an open neighborhood D of the origin in the space $\mathbb{C}^{n\times n}$ of all complex matrices Z homeomorphically onto an open neighborhood of the matrix $A \in GL(n,\mathbb{C})$; the components of the matrix Z are local coordinates in that neighborhood of A. The product space $GL(n,\mathbb{C})^{2g}$ thus has the structure of a complex analytic manifold of dimension $2gn^2$. Introduce the matrix-valued function F on the manifold $GL(n,\mathbb{C})^{2g}$ defined by

$$(3) \qquad F(S_1,\ldots,S_g,T_1,\ldots,T_g) = [S_g,T_g] \cdots [S_1,T_1] ;$$

it is obvious that this is actually a complex analytic mapping

$$F\colon GL(n,\mathbb{C})^{2g} \longrightarrow SL(n,\mathbb{C}) ,$$

where $SL(n,\mathbb{C}) \subset GL(n,\mathbb{C})$ is the subgroup consisting of matrices of determinant one, since $\det[S_j,T_j] = 1$. The subset

$$(4) \qquad R = \{(S_j,T_j) \in GL(n,\mathbb{C})^{2g} \mid F(S_j,T_j) = I\} \subset GL(n,\mathbb{C})^{2g}$$

is then a complex analytic subvariety of the complex manifold $GL(n,\mathbb{C})^{2g}$; and the mapping

$$\rho \in \text{Hom}(\pi_1(M), GL(n,\mathbb{C})) \longrightarrow (\rho(\sigma_j),\rho(\tau_j)) \in R \subset GL(n,\mathbb{C})^{2g}$$

identifies $\text{Hom}(\pi_1(M), GL(n,\mathbb{C}))$ with this subvariety, and thus establishes a complex analytic structure on the set $\text{Hom}(\pi_1(M), GL(n,\mathbf{C}))$.

Remarks. At several points, the study of Riemann surfaces inevitably leads to constructions or problems involving several complex variables. This was noted in §8 of last year's lectures, in the preliminary discussion of divisors on a Riemann surface; and it has now arisen in the discussion of flat vector bundles. So far as these lectures are concerned, really not much is involved except the definitions of a complex analytic manifold and of an analytic subvariety. (A complex analytic manifold is just the obvious generalization of a Riemann surface; an analytic subvariety is a closed subset defined locally as the set of common zeros of a finite number of complex analytic functions.) The reader should be able to follow the discussion with no further prerequisites than needed for the preceding parts of these lectures; but doubtless the discussion will be clearer to those readers having some familiarity with several complex variables.

For the case of line bundles (vector bundles of rank $n=1$), the mapping F is clearly the trivial mapping $F(S_j,T_j) \equiv 1$; thus in this case $R = GL(1,\mathbf{C})^{2g} = (\mathbf{C}^*)^{2g}$, and R itself has the structure of a complex analytic manifold of complex dimension $2g$. For the case of vector bundles of rank $n > 1$, the mapping F is non-trivial, and R is a proper subset of the complex manifold $GL(n,\mathbf{C})^{2g}$. Although R is a complex analytic subvariety, it is not a complex analytic submanifold; the subvariety R contains singularities. To investigate this situation, it is natural to introduce the differential of the mapping F . Recall that F is a complex analytic mapping between two complex analytic manifolds; for $SL(n,\mathbb{C}) \subset GL(n,\mathbb{C})$ is a Lie subgroup, hence is itself a complex manifold. The differential dF_ρ of this mapping F at a point $\rho \in GL(n,\mathbb{C})^{2g}$ is the induced linear mapping from the tangent space of $GL(n,\mathbb{C})^{2g}$ at the point ρ , to the tangent space of $SL(n,\mathbb{C})$ at the point $F(\rho)$; in more primitive terms, the differential dF_ρ is just the homogeneous linear part of the Taylor expansion of the mapping F in terms of local coordinates centered at the points ρ and $F(\rho)$. To be explicit, introduce local coordinates $(X_j,Y_j) \in (\mathbb{C}^{n\times n})^{2g}$ centered at the point $\rho = (S_j,T_j) \in GL(n,\mathbb{C})^{2g}$ by the mapping

$$(X_j,Y_j) \longrightarrow (S_j\cdot\exp X_j, T_j\cdot\exp Y_j) \ ;$$

and introduce local coordinates $Z \in \mathbb{C}^{n\times n-1}$ (viewing Z as complex matrices of trace zero) by the mapping

$$Z \longrightarrow F(\rho)\cdot\exp Z \ ,$$

recalling that $\det(\exp Z) = \exp(\operatorname{tr} Z)$. The differential $dF_\rho(X_j,Y_j)$ is just the homogeneous linear part of the Taylor expansion at the origin of the function $Z(X_j,Y_j)$ given by $F(\rho)\cdot\exp Z(X_j,Y_j) = F(S_j\cdot\exp X_j, T_j\cdot\exp Y_j)$.

Lemma 26. With the notation as above, the differential dF_ρ of the mapping F at the point $\rho = (S_j,T_j) \in GL(n,\mathbb{C})^{2g}$ is

$$(5)\quad dF_\rho(X_j,Y_j) = \sum_{j=1}^{g} \operatorname{Ad}([S_1,T_1]^{-1} \cdots [S_{j-1},T_{j-1}]^{-1}T_jS_j)\cdot \\ \cdot[(I-\operatorname{Ad}S_j^{-1})\cdot Y_j - (I-\operatorname{Ad}T_j^{-1})\cdot X_j] \ .$$

Proof. As a preliminary, recalling the Taylor series expansion of the exponential function, note that there is a Taylor expansion of the following form:

$$[S\cdot\exp X,\ T\cdot\exp Y] = S(\exp X)T(\exp Y)(\exp -X)S^{-1}(\exp -Y)T^{-1}$$
$$= [S,T] + ST\{(I - \mathrm{Ad}S^{-1})\cdot Y - (I - \mathrm{Ad}T^{-1})\cdot X\}S^{-1}T^{-1} +$$
$$+ \text{ higher powers of the variables.}$$

This leads directly to the Taylor expansion

$$F(S_j\cdot\exp X_j, T_j\cdot\exp Y_j) = [S_g\cdot\exp X_g, T_g\cdot\exp Y_g]\ldots[S_1\cdot\exp X_1, T_1\cdot\exp X_1]$$
$$= F(S_j,T_j) + \sum_{j=1}^{g}\{[S_g,T_g]\ldots[S_{j+1},T_{j+1}]S_jT_j[(I - \mathrm{Ad}\ S_j^{-1})\cdot Y_j -$$
$$- (I - \mathrm{Ad}\ T_j^{-1})\cdot X_j]S_j^{-1}T_j^{-1}[S_{j-1},T_{j-1}]\ldots[S_1,T_1]\}$$
$$+ \text{ higher powers.}$$

On the other hand, the function $Z(X_j,Y_j)$ has the Taylor expansion

$$Z(X_j,Y_j) = 0 + dF_\rho(X_j,Y_j) + \text{ higher powers;}$$

and hence

$$F(\rho)\cdot\exp Z(X_j,Y_j) = F(\rho) + F(\rho)\cdot dF_\rho(X_j,Y_j) + \text{ higher powers.}$$

Since $F(\rho)\cdot\exp Z(X_j,Y_j) = F(S_j\cdot\exp X_j,\ T_j\cdot\exp Y_j)$, the desired result follows immediately upon comparing terms, and the proof is thereby concluded.

The mapping $F\colon GL(n,\mathbb{C})^{2g} \longrightarrow SL(n,\mathbb{C})$ is <u>regular</u> at a point $\rho \in GL(n,\mathbb{C})^{2g}$ precisely when its differential dF_ρ at ρ is a linear mapping of maximal rank, that is, when $\mathrm{rank}(dF_\rho) = n^2-1$; we are of course only considering surfaces of genus $g > 1$. An open neighborhood U_ρ of a regular point ρ can be viewed as the Cartesian product and an open set in $\mathbb{C}^{(2g-1)n^2+1}$ and as an open set in $\mathbb{C}^{n^2-1}$; and the mapping F in U_ρ is just the projection onto the second factor. Letting

(6) $$R_o = \{\rho \in R \mid F \text{ is regular at } \rho\} ,$$

it follows that R_o is a complex analytic submanifold of $GL(n,\mathbb{C})^{2g}$ of complex dimension $(2g-1)n^2+1$, if R_o is non-empty. Moreover, the tangent space to the manifold R_o at a point $\rho \in R_o$ can be viewed as that vector subspace of the tangent space to $GL(n,\mathbb{C})^{2g}$ at ρ which is the kernel of the linear mapping dF_ρ . With these remarks in mind, the following is an almost immediate consequence of the preceding Lemma.

Theorem 26. The manifold R_o is the subset of the analytic variety R formed by the irreducible representations of the group $\pi_1(M)$. The tangent space to the manifold R_o at a point $\rho \in R_o$ can be identified with the space $Z^1(\pi_1(M),\mathrm{Ad}\,\rho)$ of cocycles of the group $\pi_1(M)$ with coefficients in the $\pi_1(M)$-module $\mathbb{C}^{n\times n}$ of all $n \times n$ matrices under the group representation $\mathrm{Ad}\,\rho$, (here $n > 1$).

Proof. The condition that the mapping F be regular at a point $\rho \in R$ is that $\mathrm{rank}(dF_\rho) = n^2-1$, or equivalently, that $\dim_{\mathbb{C}}(\text{kernel } dF_\rho) = (2g-1)n^2+1$, viewing the differential as a linear mapping

$$dF_\rho \colon \mathbb{C}^{2gn^2} \longrightarrow \mathbb{C}^{n^2-1} .$$

The kernel of dF_ρ is the space of all matrices $(X_j, Y_j) \in (\mathbb{C}^{n\times n})^{2g}$

such that $dF_\rho(X_j,Y_j) = 0$. However, comparing the explicit formula (5) in Lemma 26 with formulas (6) and (7) in §7, it is evident that there is a one-to-one correspondence between points (X_j,Y_j) in the kernel of dF_ρ and cocycles $(A_{\sigma_j},A_{\tau_j}) \in Z^1(\pi_1(M),\mathrm{Ad}\,\rho)$ of the group $\pi_1(M)$ with coefficients in the $\pi_1(M)$-module $\mathbb{C}^{n\times n}$ of all $n\times n$ matrices under the representation $\mathrm{Ad}\,\rho$; the correspondence is of course given by $A_{\sigma_j} = X_j$, $A_{\tau_j} = Y_j$. Thus the kernel of dF_ρ can be identified with the vector space $Z^1(\pi_1(M),\mathrm{Ad}\,\rho)$; and the condition that F be regular at ρ is just that $\dim Z^1(\pi_1(M),\mathrm{Ad}\,\rho) = (2g-1)n^2+1$. Now recall that in the proof of the Corollary to Theorem 19, on pages 133 ff., it was demonstrated that for any representation $\hat{\chi}$ of $\pi_1(M)$ of rank n , $\dim Z^1(\pi_1(M),\hat{\chi}) = (2g-1)n+q$ where q is the largest integer such that the identity representation of rank q is contained in the representation $\hat{\chi}^* = {}^t\hat{\chi}^{-1}$; since $\mathrm{Ad}\,\rho$ has rank n^2 , $\dim Z^1(\pi_1(M),\mathrm{Ad}\,\rho) = (2g-1)n^2+q$, where the identity representation of rank q is contained in $(\mathrm{Ad}\,\rho)^*$, and the condition that F be regular at ρ is just that $q = 1$. However the condition that $q = 1$ is in turn equivalent to the condition that the dimension of the vector space consisting of all complex matrices $A \in \mathbb{C}^{n\times n}$ such that $A = (\mathrm{Ad}\,\rho(\gamma))^*\cdot A = {}^t\rho(\gamma)^{-1}\cdot A\cdot {}^t\rho(\gamma)$ for all $\gamma \in \pi_1(M)$, is precisely 1 ; and this can be rewritten

$$\dim_{\mathbb{C}}\{A \in \mathbb{C}^{n\times n} \mid \rho(\gamma)\cdot {}^tA = {}^tA\cdot\rho(\gamma) \ , \ \text{all} \ \gamma \in \pi_1(M)\} = 1 \ .$$

The latter vector space always includes the space of scalar matrices, hence has dimension at least 1. If the representation ρ is irreducible, it follows from Schur's lemma that this vector space consists precisely of the scalar matrices, hence that $q = 1$; while if ρ is reducible, it is evident that this is a larger vector space, and that $q > 1$. Therefore the condition that F be regular at $\rho \in R$ is precisely that the representation ρ be irreducible; and this, together with the earlier remarks in the course of the proof, suffices to conclude the proof.

Remarks. The Jacobian matrix of the mapping F , which describes the differential dF , is evidently a complex analytic function on the manifold $GL(n,\mathbb{C})^{2g}$; so the set of those points ρ in $GL(n,\mathbb{C})^{2g}$ at which the differential dF_ρ does not attain its maximal rank is a proper analytic subvariety of $GL(n,\mathbb{C})^{2g}$. Thus R_o is the complement in R of a complex analytic subvariety of R ; indeed, clearly $R - R_o$ is the analytic subvariety consisting precisely of the singular points of the variety R . The more detailed study of the analytic space R is an interesting prospect, but would lead too far afield here. It should also be remarked that explicit form of the fundamental group is not really needed for the general results established here, but does simplify the treatment somewhat; the more general situation has been discussed elsewhere, (see the Rice University Studies (Summer 1968), Proceedings of the Complex Analysis Conference, Rice, 1967).

(b) Having shown that the set $\mathrm{Hom}(\pi_1(M), GL(n,\mathbb{C})) = R$ has a natural structure as a complex analytic variety, and that the subset of irreducible representations form a complex analytic manifold, there arises the problem of investigating the quotient spaces of these varieties modulo inner automorphisms of $GL(n,\mathbb{C})$. It is convenient to begin with a more general situation. Considering the Cartesian product manifold $GL(n,\mathbb{C})^r$ for any integers $n > 1$, $r > 1$, a point $(S_1,\dots,S_r) \in GL(n,\mathbb{C})^r$ is called <u>reducible</u> if there is a proper linear subspace of $\mathbb{C}^n$ preserved by all the linear transformations S_j ; all other points are called <u>irreducible</u>. The irreducible points form a subset $GL(n,\mathbb{C})^r_o \subset GL(n,\mathbb{C})^r$; note in passing that $R_o = R \cap GL(n,\mathbb{C})^{2g}_o$.

<u>Lemma 27.</u> For any integers $r > 1$, $n > 1$ the reducible points in $GL(n,\mathbb{C})^r$ form a proper complex analytic subvariety; hence the set $GL(n,\mathbb{C})^r_o$ of irreducible points is a dense open subset of the manifold $GL(n,\mathbb{C})^r$.

Proof. Any matrix $S \in GL(n,\mathbb{C})$ can be viewed as a complex analytic homeomorphism of the (n-1)-dimensional complex projective space $\mathbb{P}^{n-1}$, and the action of S on a point $z \in \mathbb{P}^{n-1}$ will be denoted by $S\cdot z$; the fixed points in $\mathbb{P}^{n-1}$ correspond to one-dimensional linear subspaces of $\mathbb{C}^n$ left fixed setwise by the linear transformation represented by S . In the complex analytic manifold $GL(n,\mathbb{C})^r \times \mathbb{P}^{n-1}$ consider the subset

$$V_1 = \{(S_1,\dots,S_r;z) \mid S_j\cdot z = z \quad \text{for} \quad j = 1,\dots,r\} ;$$

since the group action of the transformations is complex analytic, clearly V_1 is a complex analytic subvariety. The obvious projection mapping

$$\pi\colon GL(n,\mathbb{C})^r \times \mathbb{P}^{n-1} \longrightarrow GL(n,\mathbb{C})^r$$

is a complex analytic mapping which is proper, in the sense that the inverse image under π of any compact set is again compact; this is an immediate consequence of the observation that the manifold $\mathbb{P}^{n-1}$ is compact. Now it is known that under these circumstances the image set $\pi(V_1) \subset GL(n,\mathbb{C})^r$ is a complex analytic subvariety; this is the Remmert proper mapping theorem, (see for instance R. Gunning and H. Rossi, Analytic Functions of Several Complex Variables (Prentice-Hall, 1965), p.160; or R. Narasimhan, Introduction to the Theory of Analytic Spaces (Springer, Lecture Notes in Mathematics No.25, 1966), p.129). The variety $\pi(V_1)$ consists precisely of those points $(S_1,\ldots,S_r) \in GL(n,\mathbb{C})^r$ such that $S_j \cdot z = z$ for some point $z \in \mathbb{P}^{n-1}$ and $j = 1,\ldots,r$; thus the subset of $GL(n,\mathbb{C})^r$, consisting of r-tuples of linear transformations with one-dimensional common fixed sets, is a complex analytic subvariety. The set of all k-dimensional linear subspaces of $\mathbb{C}^n$ also has the structure of a compact complex analytic manifold for $k > 1$; these are the Grassmann manifolds $M_{n,k}$, (see for instance F. Hirzebruch, New Topological Methods in Algebraic Geometry, Springer, 1966). Repeating the above argument with the Grassmann manifold $M_{n,k}$ in place of the complex projective space $\mathbb{P}^{n-1} = M_{n,1}$, it follows that the subset of $GL(n,\mathbb{C})^r$, consisting of r-tuples of linear transformations with k-dimensional common fixed sets, is a complex analytic subvariety. Since $GL(n,\mathbb{C})^r - GL(n,\mathbb{C})^r_o$ is the union of these varieties for $k = 1,\ldots,n-1$, the proof of the lemma is completed.

<u>Remark.</u> Actually the only part of the preceding lemma which will be used is the assertion that $GL(n,\mathbb{C})^r_o$ is an open subset of the manifold $GL(n,\mathbb{C})^r$; readers wishing to avoid the machinery used in the proof of the lemma can construct a direct proof of this assertion. The fact that the reducible points form a proper subvariety is of course obvious, since it is well known that there do exist irreducible r-tuples of such matrices.

The Lie group $GL(n,\mathbb{C})$ acts as a group of complex analytic homeomorphisms of the complex manifold $GL(n,\mathbb{C})^r$ through the adjoint representation; that is, the complex analytic mapping

$$GL(n,\mathbb{C}) \times GL(n,\mathbb{C})^r \longrightarrow GL(n,\mathbb{C})^r$$

defined by

$$(T;S_1,\ldots,S_r) \longrightarrow (TS_1T^{-1},\ldots,TS_rT^{-1}) = Ad(T)\cdot(S_j) \tag{7}$$

exhibits $GL(n,\mathbb{C})$ as a complex analytic Lie transformation group acting on the complex manifold $GL(n,\mathbb{C})^r$. (For the definition and general properties of Lie transformation groups, see for example S. Helgason, <u>Differential Geometry and Symmetric Spaces</u> (Academic Press, 1962), pages 110 ff..) The center of $GL(n,\mathbb{C})$, the subgroup $Z(n,\mathbb{C}) \subset GL(n,\mathbb{C})$, consists of scalar diagonal matrices; and clearly each matrix $T \in Z(n,\mathbb{C})$ determines the identity mapping on the manifold $GL(n,\mathbb{C})^r$. Thus it is more to the point to consider the action of the quotient group $PGL(n,\mathbb{C}) = GL(n,\mathbb{C})/Z(n,\mathbb{C})$, the projective general linear group, as a transformation group on the manifold $GL(n,\mathbb{C})^r$. Note that this quotient group can also be written

$$PGL(n,\mathbb{C}) = SL(n,\mathbb{C})/SL(n,\mathbb{C}) \cap Z(n,\mathbb{C}) ,$$

since $GL(n,\mathbb{C}) = SL(n,\mathbb{C})\cdot Z(n,\mathbb{C})$. Now $SL(n,\mathbb{C}) \cap Z(n,\mathbb{C})$ consists of matrices of the form $\varepsilon\cdot I$, where I is the identity matrix and ε is a complex number such that $\varepsilon^n = 1$; this is a finite discrete subgroup of $SL(n,\mathbb{C})$, so $PGL(n,\mathbb{C})$ is a complex Lie group locally isomorphic to $SL(n,\mathbb{C})$, hence of dimension n^2-1 . The group $PGL(n,\mathbb{C})$ acts on the manifold $GL(n,\mathbb{C})^r$ in a rather complicated manner, and no attempt will be made to give a complete discussion of this matter; as usual, though, the situation can be considerably simplified by restricting attention to suitable invariant subsets of the manifold $GL(n,\mathbb{C})^r$. (Compare with the discussion in David Mumford, Geometric Invariant Theory (Springer, 1965).) The subset $GL(n,\mathbb{C})^r_o \subset GL(n,\mathbb{C})^r$ consisting of irreducible points is mapped onto itself under this group action; and in the sequel, only the group action

$$(8) \qquad PGL(n,\mathbb{C}) \times GL(n,\mathbb{C})^r_o \longrightarrow GL(n,\mathbb{C})^r_o$$

defined as in (7) will be considered. Note that each element $T \in PGL(n,\mathbb{C})$ except for the identity acts without fixed points, that is to say, that the group $PGL(n,\mathbb{C})$ acts freely on the manifold $GL(n,\mathbb{C})^r_o$; for if $T \in GL(n,\mathbb{C})$ is a matrix such that $Ad(T)\cdot S_j = TS_jT^{-1} = S_j$ for an irreducible set of matrices (S_j) , it follows from Schur's lemma that $T \in Z(n,\mathbb{C})$.

Lemma 28. For any integers $r > 1$, $n > 1$, each point $(S_j) \in GL(n,\mathbb{C})^r_o$ has an open neighborhood U such that the set

$$\{T \in PGL(n,\mathbb{C}) | Ad(T)\cdot U \cap U \neq \emptyset\}$$

has compact closure in $PGL(n,\mathbb{C})$.

Proof. Recall from Lemma 27 that $GL(n,\mathbb{C})^r_o$ is an open subset of the manifold $GL(n,\mathbb{C})^r$. Each point $(S_j) \in GL(n,\mathbb{C})^r_o$ has an open neighborhood U such that the point set closure $\overline{U}$ is compact and $\overline{U} \subset GL(n,\mathbb{C})^r_o$. Note that the unitary matrices form a compact subgroup $U(n) \subset GL(n,\mathbb{C})$; so replacing the set U by $Ad(U(n))\cdot U$ if necessary, there is no loss of generality in assuming further that the neighborhood U is mapped onto itself by the subgroup $U(n) \subset GL(n,\mathbb{C})$ of unitary matrices. This neighborhood U then has the desired properties. For suppose that, in contradiction of the desired result, there exists a sequence of elements $T_\nu \in PGL(n,\mathbb{C})$ such that $Ad(T_\nu)\cdot U \cap U \neq \emptyset$ for each ν , but that the elements T_ν have no limit point in $PGL(n,\mathbb{C})$; of course, the elements T_ν can be viewed as matrices $T_\nu \in SL(n,\mathbb{C})$ rather than as elements of the quotient group $PGL(n,\mathbb{C})$, with no change in the statements. Each matrix T_ν can be written in the form $T_\nu = A_\nu\cdot D_\nu\cdot B_\nu$, where A_ν, B_ν are unitary matrices and D_ν is a diagonal matrix of determinant one, say $D_\nu = \text{diag}(d_1^\nu,\ldots,d_n^\nu)$. Now on the one hand, the matrices D_ν will have no limit points in $SL(n,\mathbb{C})$ either; so after passing to a subsequence if necessary, and relabeling the matrices, it can be assumed that $|d_i^\nu| \longrightarrow \infty$ as $\nu \longrightarrow \infty$ for $i = 1,\ldots,m$, but that $|d_i^\nu|$ remain bounded as $\nu \longrightarrow \infty$ for $i = m+1,\ldots,n$. Since $d_1^\nu\cdot\ldots\cdot d_n^\nu = 1$, it necessarily follows that $1 \leq m < n$. On the other hand, $Ad(D_\nu)\cdot U \cap U \neq \emptyset$ for each ν , since the set U is mapped onto itself by any unitary matrix; therefore there are points $(S_j^\nu) \in U$ so that $\mathrm{Ad}(D_\nu)\cdot S_j^\nu \in U$ for each index ν. Since $\overline{U}$ is compact, after passing to a further subsequence if necessary, it can be assumed that the matrices (S_j^ν) converge to a point $(S_j) \in \overline{U} \subset GL(n,\mathbb{C})_o^r$ as $\nu \longrightarrow \infty$. Writing these matrices out explicitly as $(S_j^\nu) = (s_{k\ell}^{\nu j})$, $(S_j) = (s_{k\ell}^j) = \lim_{\nu \to \infty} (s_{k\ell}^{\nu j})$, note that

$$\mathrm{Ad}(D_\nu)\cdot(S_j^\nu) = \frac{d_k^\nu}{d_\ell^\nu}\, s_{k\ell}^{\nu j} .$$

Since the points (S_j^ν) belong to the compact set $\overline{U}$, all the components of these matrices are bounded; that is,

$$\left|\frac{d_k^\nu}{d_\ell^\nu}\, s_{k\ell}^{\nu j}\right| \leq M < \infty \quad \text{for all} \quad \nu, j, k, \ell .$$

Letting $\nu \longrightarrow \infty$, note that $|d_k^\nu / d_\ell^\nu| \longrightarrow \infty$ whenever $k = 1,\ldots,m$ and $\ell = m+1,\ldots,n$; and therefore $s_{k\ell}^j = \lim_{\nu \to \infty} s_{k\ell}^{\nu j} = 0$ whenever $k = 1,\ldots,m$ and $\ell = m+1,\ldots,n$. This means that the point (S_j) is reducible, which is impossible since $(S_j) \in \overline{U} \subset GL(n,\mathbb{C})_o^r$; and this contradiction serves to conclude the proof.

Theorem 27. For any integers $r > 1$, $n > 1$, the quotient space

$$H(n,\mathbb{C})_o^r = GL(n,\mathbb{C})_o^r / PGL(n,\mathbb{C})$$

under the group action (7) has the structure of a complex analytic manifold such that the natural projection

$$\pi\colon GL(n,\mathbb{C})_o^r \longrightarrow H(n,\mathbb{C})_o^r$$

is a complex analytic principal $PGL(n,\mathbb{C})$ bundle.

Remarks. The conclusion of the theorem is just that the quotient space $H(n,\mathbb{C})^r_o$, with the natural quotient space topology, can be given the structure of a complex analytic manifold, in such a manner that each point of $H(n,\mathbb{C})^r_o$ has an open neighborhood V for which the inverse image $\pi^{-1}(V) \subset GL(n,\mathbb{C})^r_o$ is analytically homeomorphic to the product manifold $PGL(n,\mathbb{C}) \times V$; and the homeomorphism $\pi^{-1}(V) \cong PGL(n,\mathbb{C}) \times V$ commutes with the obvious actions of the group $PGL(n,\mathbb{C})$ on the two sets. For a general discussion of such group actions, see Richard S. Palais, "On the existence of slices for actions of non-compact Lie groups," Annals of Math. 73 (1961), 295-323.

Proof. Select a fixed point $(S_j) \in GL(n,\mathbb{C})^r_o$, and consider the complex analytic mapping

$$G\colon PGL(n,\mathbb{C}) \longrightarrow GL(n,\mathbb{C})^r_o$$

defined by $G(T) = Ad(T)\cdot(S_j)$ for any $T \in PGL(n,\mathbb{C})$. Since the group $PGL(n,\mathbb{C})$ acts freely, this is a regular mapping at each point of $PGL(n,\mathbb{C})$, in the sense that the differential dG_T has maximal rank at each point $T \in PGL(n,\mathbb{C})$. This is true quite generally; but for the sake of completeness, and since the explicit formulas will be needed later anyway, a special proof will be given for this instance of the result. For the result in general, see for instance L. P. Eisenhart, Continuous Groups of Transformations (Dover, 1961; Princeton University Press, 1933), especially Chapter I. Introduce local coordinates $(X_j) \in (\mathbb{C}^{n\times n})^r$ centered at the point $G(T) = (TS_jT^{-1})$, by the mapping $(X_j) \longrightarrow (TS_jT^{-1}\cdot \exp X_j)$; and introduce local coordinates $Z \in \mathbb{C}^{n\times n-1}$ (viewing Z as complex matrices of trace zero) centered at the point $T \in PGL(n,\mathbb{C})$, by the mapping $Z \longrightarrow T\cdot\exp Z$. The group $PGL(n,\mathbb{C})$ is again locally identified with the group $SL(n,\mathbb{C})$. In terms of these local coordinates, the mapping G is described by the coordinate functions $(G_j(Z)) \in (\mathbb{C}^{n\times n})^r$, where

$$TS_jT^{-1}\cdot\exp G_j(Z) = (T\exp Z)S_j(T\exp Z)^{-1} ;$$

expanding both sides of this equality in a Taylor series in the variable Z , note that

$$TS_jT^{-1}(I + G_j(Z) + \text{higher powers of } G_j(Z)) = \\ = T(I + Z + \ldots)S_j(I - Z + \ldots)T^{-1} .$$

The differential $dG_T(Z)$ is just the homogeneous linear part of the Taylor expansion of the function $(G_j(Z))$, so that

$$(9) \qquad \begin{cases} dG_T(Z) = T(S_j^{-1}ZS_j - Z)T^{-1} \\ \qquad\qquad = Ad(T)\cdot[(Ad(S_j^{-1}) - I)\cdot Z] . \end{cases}$$

To show that the linear mapping dG_T is of maximal rank, it suffices to show that it has trivial kernel. If Z is a matrix such that $dG_T(Z) = 0$, it follows from (9) that $Z = Ad(S_j^{-1})\cdot Z = S_j^{-1}ZS_j$. Since (S_j) is an irreducible point, Schur's lemma shows that Z is a scalar matrix; and since $\operatorname{tr} Z = 0$, necessarily $Z = 0$, and so dG_T has trivial kernel as desired.

Now since the mapping G is regular, there is an open neighborhood Δ of the identity $I \in PGL(n,\mathbb{C})$ such that G is a complex analytic homeomorphism between Δ and a complex analytic submanifold $G(\Delta)$ of an open neighborhood U of (S_j) in $GL(n,\mathbf{C})$. Select an analytic submanifold $V \subset U$ such that $(S_j) \in V$, and that the tangent spaces of the submanifolds $G(\Delta)$ and V at the point (S_j) are linearly independent subspaces which span the full tangent space of the manifold $GL(n,\mathbf{C})^r_o$. The complex analytic mapping $\Delta \times V \longrightarrow GL(n,\mathbf{C})^r_o$, defined by $(T,(X_j)) \longrightarrow Ad(T)\cdot(X_j)$ for $T \in \Delta \subset PGL(n,\mathbf{C})$ and $(X_j) \in V \subset GL(n,\mathbf{C})^r_o$, is then also a regular mapping at the point $(I,(S_j))$; hence, after restricting the neighborhoods suitably, this mapping is a complex analytic homeomorphism $\Delta \times V \cong U$. To complete the proof of the theorem, it is only necessary to show that this mapping extends to a complex analytic homeomorphism from $PGL(n,\mathbf{C}) \times V$ into $GL(n,\mathbf{C})^r_o$; and it is clear that for this purpose it suffices to show that, after restricting the neighborhoods further if necessary, no two points of the submanifold V are equivalent under the action of the group $PGL(n,\mathbf{C})$. Suppose, contrariwise, that there are sequences of points $(X'_{j\nu})$, $(X''_{j\nu}) \in V$, and of transformations $T_\nu \in PGL(n,\mathbf{C})$, such that $\lim_{\nu\to\infty}(X'_{j\nu}) = \lim_{\nu\to\infty}(X''_{j\nu}) = (S_j)$ and $Ad(T_\nu)\cdot(X'_{j\nu}) = (X''_{j\nu})$ for each ν. From Lemma 28 it is clear that upon choosing a suitable subsequence, the transformations T_ν can be assumed to converge to some element $T \in PGL(n,\mathbf{C})$; and since $Ad(T)\cdot(S_j) = \lim_{\nu\to\infty} Ad(T_\nu)\cdot(X'_{j\nu}) = (S_j)$, necessarily T is the identity transformation. However this means that $T_\nu \in \Delta$ for ν sufficiently large, and it is impossible that $Ad(T_\nu)\cdot(X'_{j\nu}) = (X''_{j\nu}) = Ad(I)\cdot(X''_{j\nu})$ since the mapping is a homeomorphism from $\Delta \times V$

onto U. This contradiction serves to complete the proof of the theorem.

To apply this theorem to the particular case at hand, consider again the complex analytic submanifold

$$R_o = R \cap GL(n,\mathbf{C})_o^{2g} \subset GL(n,\mathbf{C})_o^{2g}$$

defined by (4). For any element $P \in PGL(n,\mathbf{C})$ and any point $(S_j,T_j) \in GL(n,\mathbf{C})^{2g}$, note that $F(Ad(P)\cdot(S_j,T_j)) = Ad(P)\cdot F(S_j,T_j)$; it is thus evident that the submanifold $R_o \subset GL(n,\mathbf{C})_o^{2g}$ is preserved by the adjoint action of the group $PGL(n,\mathbf{C})$, and that the quotient space

$$(10) \qquad S_o = R_o/PGL(n,\mathbf{C}) \subset GL(n,\mathbf{C})_o^{2g}/PGL(n,\mathbf{C}) = H(n,\mathbf{C})_o^{2g}$$

is a complex analytic submanifold of the complex manifold $H(n,\mathbf{C})_o^{2g}$. Indeed, R_o has the inherited structure of a complex analytic principal $PGL(n,\mathbf{C})$ bundle over the manifold S_o. Note in passing that $\dim H(n,\mathbf{C})_o^{2g} = \dim GL(n,\mathbf{C})_o^{2g} - \dim PGL(n,\mathbf{C}) = (2g-1)n^2+1$, and that $\dim S_o = \dim R_o - \dim PGL(n,\mathbf{C}) = 2(g-1)n^2+2$. In summary, the following holds:

<u>Theorem 28.</u> For $n > 1$, the quotient space $S_o = R_o/PGL(n,C)$ has the structure of a complex analytic manifold such that the natural projection $R_o \longrightarrow S_o$ is a complex analytic principal $PGL(n,C)$ bundle. The tangent space to the manifold S_o at a point $\rho \in S_o$ can be identified with the cohomology group $H^1(\pi_1(M),Ad\,\rho)$ of the group $\pi_1(M)$ with coefficients in the $\pi_1(M)$-module $\mathbf{C}^{n\times n}$ of all $n \times n$ matrices under the group representation $Ad\,\rho$.

Proof. The first assertion, as noted above, is an almost immediate corollary of Theorem 27. For the second assertion, recall from Theorem 26 that the tangent space to the manifold R_o at a point $\rho \in R_o$ can be identified with the cocycle group $Z^1(\pi_1(M), \mathrm{Ad}\,\rho)$; indeed, using local coordinates to identify the tangent space to the manifold $GL(n,\mathbf{C})_o^{2g}$ at the point ρ with the vector space $(\mathbf{C}^{n\times n})^{2g}$ of 2g-tuples of complex matrices, the tangent space to the submanifold $R_o \subset GL(n,\mathbf{C})_o^{2g}$ is the subspace consisting of those 2g-tuples of matrices (X_j, Y_j) such that $A_{\sigma_j} = X_j$, $A_{\tau_j} = Y_j$, is a cocycle $A \in Z^1(\pi_1(M), \mathrm{Ad}\,\rho)$. The tangent space to the quotient manifold $S_o = R_o/PGL(n,\mathbf{C})$ at a point corresponding to a representation $\rho \in R_o$ is just the quotient space of $Z^1(\pi_1(M), \mathrm{Ad}\,\rho)$ modulo the vector subspace which is the tangent space to the orbit $\mathrm{Ad}(PGL(n,\mathbf{C}))\cdot\rho$ at the point ρ . The tangent spaces to the orbits were essentially calculated in the course of the proof of Theorem 27, however, recalling formula (9); the tangent space to the orbit at ρ is just the vector space of matrices $(X_j, Y_j) \in (\mathbf{C}^{n\times n})^{2g}$ of the form

$$X_j = (\mathrm{Ad}(S_j^{-1}) - I)\cdot Z , \qquad Y_j = (\mathrm{Ad}(T_j^{-1}) - I)\cdot Z ,$$

where $\rho = (S_j, T_j)$ and Z are arbitrary complex matrices of trace zero. Recalling formula (8) of §7, these matrices correspond precisely to the group $B^1(\pi_1(M), \mathrm{Ad}\,\rho)$ of one-coboundaries; and therefore the tangent space to S_o at ρ is precisely $Z^1(\pi_1(M), \mathrm{Ad}\,\rho)/B^1(\pi_1(M), \mathrm{Ad}\,\rho) = H^1(\pi_1(M), \mathrm{Ad}\,\rho)$, which serves to complete the proof.

This theorem then establishes a natural structure of a complex analytic manifold of dimension $2(g-1)n^2+2$ on the set $H^1(M,GL(n,\mathbb{C}))_0 \subset H^1(M,GL(n,\mathbb{C}))$ of irreducible flat complex vector bundles of rank $n > 1$ over the compact Riemann surface M ; the same assertion holds for the case $n = 1$, since all flat line bundles over M are irreducible, and the set of all such has the natural structure of the complex manifold $(\mathbb{C}^*)^{2g}$ of dimension $2g$.

(c) On the quotient manifold $S_0 = R_0/PGL(n,\mathbb{C})$ there is a further equivalence relation to be investigated, the complex analytic equivalence of flat vector bundles. Recall that in the case of bundles of rank $n = 1$, the analytically trivial flat line bundles form the Lie subgroup $\delta\Gamma(M,\mathcal{O}^{1,0}) \subset H^1(M,\mathbf{C}^*)$; and the set of analytic equivalence classes of flat line bundles is the quotient Lie group $H^1(M,\mathbf{C}^*)/\delta\Gamma(M,\mathcal{O}^{1,0})$, which is the Picard variety of the Riemann surface M , (recalling §8 of last year's lectures). The set of flat vector bundles of rank $n > 1$ is not a group, so the investigation of the corresponding space of equivalence classes is rather more complicated. We shall begin by examining individual equivalence classes, as represented by complex analytic connections.

Let $\mathfrak{U} = \{U_\alpha\}$ be an open covering of the compact Riemann surface M , such that $\mathfrak{U}$ is a Leray covering for both flat sheaves and analytic sheaves. For a given flat vector bundle $\chi \in H^1(M,GL(n,\mathbf{C}))$ select a representative cocycle $(\chi_{\alpha\beta}) \in Z^1(\mathfrak{U},GL(n,\mathbf{C}))$. Recall from Theorem 17 in §6(c) that there is a natural one-to-one correspondence between the set of flat vector bundles analytically equivalent to χ and the set $\Lambda^{*}(\mathfrak{U},\chi)$ of equivalence classes of complex analytic connections for the bundle χ. Explicitly, as in §6(c) again, a connection $\lambda \in \Lambda(\mathfrak{U},\chi)$ is a zero-cochain $(\lambda_{\alpha}) \in C^{0}(\mathfrak{U}, \mathcal{O}^{1,0}(\mathrm{Ad}\ \chi))$ such that $\delta\lambda = D\chi$; since χ is a flat bundle, $D\chi = 0$ and hence $\delta\lambda = 0$. Therefore for a flat vector bundle χ it follows that $\Lambda(\mathfrak{U},\chi) = Z^{0}(\mathfrak{U}, \mathcal{O}^{1,0}(\mathrm{Ad}\ \chi)) = \Gamma(M,\mathcal{O}^{1,0}(\mathrm{Ad}\ \chi))$; or equivalently, the complex analytic connections for a flat vector bundle χ can be identified with the space of Prym differentials for the bundle $\mathrm{Ad}\ \chi$. Now to each complex analytic connection or Prym differential $\lambda \in \Gamma(M,\mathcal{O}^{1,0}(\mathrm{Ad}\ \chi))$ there is canonically associated a flat vector bundle $H_{\chi}(\lambda) \in H^{1}(M,GL(n,\mathbb{C}))$ analytically equivalent to χ, as follows. Select holomorphic functions $F_{\alpha} \in GL(n,\mathcal{O}_{U_{\alpha}})$ in the various sets U_{α} such that $DF_{\alpha} = \lambda_{\alpha}$, and set $\chi'_{\alpha\beta} = F_{\alpha}\cdot\chi_{\alpha\beta}\cdot F_{\beta}^{-1}$ in $U_{\alpha} \cap U_{\beta}$; the cocycle $(\chi'_{\alpha\beta}) \in Z^{1}(\mathfrak{U},GL(n,\mathbb{C}))$ represents the flat vector bundle $H_{\chi}(\lambda)$. There is thus a well-defined mapping

$$H_{\chi}\colon \Gamma(M,\mathcal{O}^{1,0}(\mathrm{Ad}\ \chi)) \longrightarrow H^{1}(M,GL(n,\mathbb{C})), \tag{11}$$

such that the image of H_{χ} consists precisely of the flat vector bundles analytically equivalent to the given bundle χ; note that in particular, $H_{\chi}(0) = \chi$, where 0 denotes the trivial Prym differential.

If $\chi' = H_{\chi}(\lambda') \in H^{1}(M,GL(n,\mathbb{C}))$ for some connection $\lambda' \in \Gamma(M,\mathcal{O}^{1,0}(\mathrm{Ad}\ \chi))$, there is a corresponding mapping

$$H_{\chi'}\colon \Gamma(M,\mathcal{O}^{1,0}(\mathrm{Ad}\ \chi')) \longrightarrow H^{1}(M,GL(n,\mathbb{C})),$$

and the mappings H_χ and $H_{\chi'}$ have precisely the same images; the relation between these two mappings can be described as follows. Select holomorphic functions $G_\alpha \in GL(n, \mathcal{O}_{U_\alpha})$ in the various sets U_α such that $DG_\alpha = \lambda'_\alpha$; thus the cocycle

$$(\chi'_{\alpha\beta}) = (G_\alpha \cdot \chi_{\alpha\beta} \cdot G_\beta^{-1}) \in Z^1(\mathcal{U}, GL(n,\mathbb{C}))$$

represents the flat vector bundle χ' . The mapping

$$G\colon \Gamma(M, \mathcal{O}^{1,0}(\mathrm{Ad}\ \chi)) \longrightarrow \Gamma(M, \mathcal{O}^{1,0}(\mathrm{Ad}\ \chi'))$$

defined by

(12) $(G\lambda)_\alpha = \mathrm{Ad}(G_\alpha)\cdot(\lambda_\alpha - \lambda'_\alpha)$, for any $\lambda = (\lambda_\alpha) \in \Gamma(M, \mathcal{O}^{1,0}(\mathrm{Ad}\ \chi))$

is clearly a non-singular affine mapping from the space $\Gamma(M, \mathcal{O}^{1,0}(\mathrm{Ad}\ \chi))$ onto the space $\Gamma(M, \mathcal{O}^{1,0}(\mathrm{Ad}\ \chi'))$; and moreover,

(13) $$H_\chi = H_{\chi'} \circ G .$$

To see this, note that for any connection $\lambda = (\lambda_\alpha)$ the image bundle $H_\chi(\lambda)$ is represented by the cocycle $(F_\alpha \cdot \chi_{\alpha\beta} \cdot F_\beta^{-1})$, where $F_\alpha \in GL(n, \mathcal{O}_{U_\alpha})$ are holomorphic functions such that $DF_\alpha = \lambda_\alpha$. The same cocycle can also be written $(F_\alpha G_\alpha^{-1} \cdot \chi'_{\alpha\beta} \cdot G_\beta F_\beta^{-1})$; and since $D(F_\alpha G_\alpha^{-1}) = \mathrm{Ad}(G_\alpha)\cdot(DF_\alpha - DG_\alpha) = (G\lambda)_\alpha$, it follows that $H_\chi(\lambda) = H_{\chi'}(G\lambda)$, as desired. Since the principal interest here lies in the image of the mapping H_χ , rather than in the mapping itself, the relation (13) is useful for obtaining a number of different representations for that image set. In particular, it is clear that for studying the local properties of the mapping, it suffices to restrict attention to neighborhoods of the zero element of $\Gamma(M, \mathcal{O}^{1,0}(\text{Ad } X))$.

Returning to the mapping (11) again, note that $\Gamma(M, \mathcal{O}^{1,0}(\text{Ad } X))$ is a finite-dimensional complex vector space, hence has the natural structure of a complex analytic manifold. Recalling Theorem 28, the subset $H^1_o(M, GL(n, \mathbb{C})) \subset H^1(M, GL(n, \mathbb{C}))$ consisting of irreducible flat vector bundles also has the structure of a complex analytic manifold, indeed, can be identified with the manifold S_o described above. Letting $\Gamma_o(M, \mathcal{O}^{1,0}(\text{Ad } X)) \subset \Gamma(M, \mathcal{O}^{1,0}(\text{Ad } X))$ be the subset consisting of those complex analytic connections which determine irreducible flat vector bundles, the restriction of the mapping (11) is the mapping

$$H^o_X \colon \Gamma_o(M, \mathcal{O}^{1,0}(\text{Ad } X)) \longrightarrow H^1_o(M, GL(n, \mathbb{C})) = S_o$$

from a subset of the complex manifold $\Gamma(M, \mathcal{O}^{1,0}(\text{Ad } X))$ into the complex manifold S_o .

<u>Lemma 29.</u> The subset $\Gamma_o(M, \mathcal{O}^{1,0}(\text{Ad } X)) \subset \Gamma(M, \mathcal{O}^{1,0}(\text{Ad } X))$ is the complement of a complex analytic subvariety of $\Gamma(M, \mathcal{O}^{1,0}(\text{Ad } X))$, hence is itself a connected complex analytic manifold; and the mapping H^o_X is a complex analytic mapping.

Proof. Select any point $\lambda^o \in \Gamma(M, \mathcal{O}^{1,0}(\text{Ad } X))$, and let $\lambda^i \in \Gamma(M, \mathcal{O}^{1,0}(\text{Ad } X))$, $1 \leq i \leq r$, be a basis for this space of Prym differentials; then local complex analytic coordinates $t = (t_1, \ldots, t_r)$ centered at λ^o can be introduced by the mapping

$$(t_1, \ldots, t_r) \longrightarrow (\lambda^o + \sum_{i=1}^{r} t_i \lambda^i) .$$

In each coordinate neighborhood $U_\alpha \subset M$ with local coordinate z_α, select a point p_α; and let $\Delta \subset \mathbf{C}^r$ be an open neighborhood of the origin in the space $\mathbf{C}^r$ with coordinates $(t_1,\ldots,t_r)$. In the open subset $U_\alpha \times \Delta \subset \mathbf{C}^{1+r}$ consider the system of partial differential equations

$$(14) \qquad \frac{\partial}{\partial z_\alpha} F_\alpha(z_\alpha,t) = F_\alpha(z_\alpha,t)\cdot[\lambda^0_\alpha(z_\alpha) + \sum_{i=1}^{r} t_i\lambda^i_\alpha(z_\alpha)] ,$$

where $F_\alpha \colon U_\alpha \times \Delta \longrightarrow \mathbf{C}^{n\times n}$ are matrix-valued functions subject to the initial conditions

$$(15) \qquad F_\alpha(p_\alpha,t) \equiv I \quad \text{for all} \quad t \in \Delta .$$

It follows from the Cauchy-Kowalewsky Theorem (see for instance Courant-Hilbert, Methods of Mathematical Physics (Interscience, 1962), vol. II), that there are unique holomorphic solutions $F_\alpha(z_\alpha,t)$ of the differential equation (14) satisfying the initial conditions (15), provided that the neighborhoods U_α and Δ both are chosen sufficiently small; since the covering $\mathfrak{U}$ can be assumed finite, M being compact, the same set Δ can be used with all the sets U_α. The function F_α can of course be assumed to be non-singular throughout $U_\alpha \times \Delta$. For any fixed point $t = (t_1,\ldots,t_r) \in \Delta$, the cocycle

$$(\chi_{\alpha\beta}(t)) = (F_\alpha(z_\alpha,t)\cdot\chi_{\alpha\beta}\cdot F_\beta(z_\beta,t)) \in Z^1(\mathfrak{U},GL(n,\mathbf{C}))$$

represents the flat vector bundle $H_\chi(\lambda^0 + \sum_{i=1}^{r} t_i\lambda^i)$; and the components of the matrices $(\chi_{\alpha\beta}(t))$ are holomorphic functions of t in the set Δ. Passing to the characteristic representation of the bundle, it is obvious that there results a complex analytic mapping

$$(16) \qquad \Delta \longrightarrow R = \mathrm{Hom}(\pi_1(M), GL(n,\mathbb{C})) \subset GL(n,\mathbb{C})^{2g} .$$

The subset $R_o \subset R$ of irreducible representations is the complement of a complex analytic subvariety of R, as seen earlier; so, since the mapping (16) is complex analytic, the inverse image Δ_o of the set R_o is either empty, or all of Δ, or the complement of a proper complex analytic subvariety of Δ. Note that this proves the first assertion of the lemma. Upon restricting the mapping (16) to the subset $\Delta_o \subset \Delta$, and following it with the complex analytic projection $R_o \longrightarrow R_o/PGL(n,\mathbb{C}) = S_o$, it follows that the mapping H_χ^o is a complex analytic mapping in the set Δ_o, and the proof is thereby concluded.

Remarks. If the given bundle χ is not irreducible, the subset $\Gamma_o(M, \mathcal{O}^{1,0}(\mathrm{Ad}\ \chi)) \subset \Gamma(M, \mathcal{O}^{1,0}(\mathrm{Ad}\ \chi))$ may be empty, insofar as the preceding lemma goes; this point will be taken up again later. Of course, in view of the remarks immediately preceding the lemma, it would have been sufficient to take $\lambda^o = 0$; for the mapping (12) is clearly complex analytic. This was not done, merely because the saving in effort would have been negligible.

The mapping H_χ of (11) is not always a one-to-one mapping onto its image; as proved in Theorem 17, two complex analytic connections have the same image under H_χ if and only if they are equivalent in the sense of the definition on page 115. In order to study the image of H_χ, it is convenient to pass to the set of equivalence classes of connections, and to consider the mapping induced by H_X on this set; and that requires a more detailed examination of the equivalence relation.

For any flat vector bundle $X \in H^1(M, GL(n, \mathbb{C}))$, the set $\mathrm{End}(X)$ of complex analytic endomorphisms of X is a finite dimensional complex vector space, which can be identified with the space $\Gamma(M, \mathcal{O}(\mathrm{Ad}\ X))$; in terms of an open covering $\mathfrak{U} = \{U_\alpha\}$ of the surface M, an element $T \in \mathrm{End}(X)$ is described by holomorphic matrix-valued functions $T_\alpha \in \mathcal{O}^{n \times n}_{U_\alpha}$ such that $T_\alpha X_{\alpha\beta} = X_{\alpha\beta} T_\beta$, or equivalently $T_\alpha = \mathrm{Ad}(X_{\alpha\beta}) \cdot T_\beta$, in each intersection $U_\alpha \cap U_\beta$. Actually of course, $\mathrm{End}(X)$ is a finite dimensional algebra over the complex numbers, as discussed on page 105. The set of invertible endomorphisms is the group $\mathrm{Aut}(X) \subset \mathrm{End}(X)$ of complex analytic automorphisms of X ; clearly $\mathrm{Aut}(X)$ is a complex Lie group of dimension s, where $s = \dim \Gamma(M, \mathcal{O}(\mathrm{Ad}\ X))$. The group $\mathrm{Aut}(X)$ contains as a Lie subgroup the set of automorphisms of the form $T_\alpha = cI$, where $I \in GL(n, \mathbb{C})$ is the identity matrix and $c \in \mathbb{C}^*$ is an arbitrary non-zero constant; this is a normal, even a central, subgroup of $\mathrm{Aut}(X)$, and is isomorphic to $\mathbb{C}^*$. The quotient group $\mathrm{Aut}(X)/\mathbb{C}^* = P\,\mathrm{Aut}(X)$ will be called the projective group of complex analytic automorphisms of the bundle X ; note that $P\,\mathrm{Aut}(X)$ is a complex Lie group of dimension $s - 1$. As on page 115, to each $T \in \mathrm{Aut}(X)$ there is associated the mapping

$$\mathrm{Ad}^*(T) \colon \Gamma(M, \mathcal{O}^{1,0}(\mathrm{Ad}\ X)) \longrightarrow \Gamma(M, \mathcal{O}^{1,0}(\mathrm{Ad}\ X))$$

defined by

$$(17) \qquad (\mathrm{Ad}^*(T) \cdot \lambda)_\alpha = \mathrm{Ad}(T_\alpha) \cdot (\lambda_\alpha - DT_\alpha) = \mathrm{Ad}(T_\alpha) \cdot \lambda_\alpha - dT_\alpha \cdot T_\alpha^{-1}$$

for any $\lambda = (\lambda_\alpha) \in \Gamma(M, \mathcal{O}^{1,0}(\mathrm{Ad}\ X))$. It is easily seen that this exhibits $\mathrm{Aut}(X)$ as a complex Lie group of nonsingular complex affine transformations of the vector space $\Gamma(M, \mathcal{O}^{1,0}(\mathrm{Ad}\ X))$; actually, since it is obvious that the automorphisms of the form $T_\alpha = cI$ act trivially, this also defines a similar action of the quotient group $P\ \mathrm{Aut}(X)$. The quotient space

$$\Gamma(M, \mathcal{O}^{1,0}(\mathrm{Ad}\ X))/\ P\ \mathrm{Aut}(X)$$

under this group action is precisely the space $\Lambda^*(\mathfrak{M}, X)$ of equivalence classes of complex analytic connections for the vector bundle X .

As a slight digression, consider an equivalent bundle $X' = H_X(\lambda')$, for some connection $\lambda' \in \Gamma(M, \mathcal{O}^{1,0}(\mathrm{Ad}\ X))$; and as before introduce the mapping

$$G\colon \Gamma(M, \mathcal{O}^{1,0}(\mathrm{Ad}\ X)) \longrightarrow \Gamma(M, \mathcal{O}^{1,0}(\mathrm{Ad}\ X'))$$

defined by (12), for some functions $G_\alpha \in GL(n, \mathcal{O}_{U_\alpha})$ such that $DG_\alpha = \lambda'_\alpha$. There is a corresponding mapping

$$G\colon \Gamma(M, \mathcal{O}(\mathrm{Ad}\ X)) \longrightarrow \Gamma(M, \mathcal{O}(\mathrm{Ad}\ X'))$$

defined by

(19) $\quad (GT)_\alpha = \mathrm{Ad}(G_\alpha)\cdot T_\alpha$, for any $T = (T_\alpha) \in \Gamma(M, \mathcal{O}(\mathrm{Ad}\ X))$;

and this of course induces a mapping

$$G\colon \mathrm{Aut}(X) \longrightarrow \mathrm{Aut}(X') .$$

Now it is easy to see that

(20) $$G\cdot \mathrm{Ad}^*(T) = \mathrm{Ad}^*(GT)\cdot G ,$$

or equivalently, that the following diagram is commutative for any $T \in \mathrm{Aut}(X)$:

$$\begin{array}{ccc} \Gamma(M, \mathcal{O}^{1,0}(\mathrm{Ad}\ X)) & \xrightarrow{\mathrm{Ad}^*(T)} & \Gamma(M, \mathcal{O}^{1,0}(\mathrm{Ad}\ X)) \\ \downarrow G & & \downarrow G \\ \Gamma(M, \mathcal{O}^{1,0}(\mathrm{Ad}\ X')) & \xrightarrow{\mathrm{Ad}^*(GT)} & \Gamma(M, \mathcal{O}^{1,0}(\mathrm{Ad}\ X')) \ . \end{array}$$

For given any connection $\lambda = (\lambda_\alpha) \in \Gamma(M, \mathcal{O}^{1,0}(\mathrm{Ad}\ X))$, it follows that

$$\mathrm{Ad}^*(GT)\cdot G\cdot\lambda = \mathrm{Ad}(G_\alpha T_\alpha G_\alpha^{-1})\cdot[\mathrm{Ad}(G_\alpha)\cdot(\lambda_\alpha - \lambda_\alpha') - D(G_\alpha T_\alpha G_\alpha^{-1})]$$

$$= \mathrm{Ad}(G_\alpha T_\alpha G_\alpha^{-1})\cdot[\mathrm{Ad}(G_\alpha)\cdot(\lambda_\alpha - DG_\alpha) - \mathrm{Ad}(G_\alpha T_\alpha^{-1})\cdot DG_\alpha - \mathrm{Ad}(G_\alpha)\cdot DT_\alpha + \mathrm{Ad}(G_\alpha)\cdot DG_\alpha]$$

$$= \mathrm{Ad}(G_\alpha)\cdot[\mathrm{Ad}(T_\alpha)\cdot(\lambda_\alpha - DT_\alpha) - \lambda_\alpha'] = G\cdot\mathrm{Ad}^*(T)\cdot\lambda \ .$$

Thus the mappings G transform the action of the transformation group $\mathrm{Aut}(X)$ on the space $\Gamma(M, \mathcal{O}^{1,0}(\mathrm{Ad}\ X))$ into the action of the transformation group $\mathrm{Aut}(X')$ on the space $\Gamma(M, \mathcal{O}^{1,0}(\mathrm{Ad}\ X'))$, whenever X' is equivalent to X ; once again, this is a useful observation, enabling local questions to be considered near the neighborhood of the trivial connection $\lambda = 0$ alone.

Now the question arises how this action of the transformation group $\mathrm{Aut}(X)$ affects the special subset $\Gamma_0(M, \mathcal{O}^{1,0}(\mathrm{Ad}\ X)) \subset \Gamma(M, \mathcal{O}^{1,0}(\mathrm{Ad}\ X))$, as above; and the answer is provided by the following simple but interesting observation.

Lemma 30. A connection $\lambda \in \Gamma(M, \mathcal{O}^{1,0}(\mathrm{Ad}\ X))$ is left fixed by a transformation $T \in P\ \mathrm{Aut}(X)$, $T \neq I$, if and only if

λ corresponds to a reducible flat vector bundle. The subset $\Gamma_0(M, \mathcal{O}^{1,0}(\mathrm{Ad}\ X)) \subset \Gamma(M, \mathcal{O}^{1,0}(\mathrm{Ad}\ X))$ of connections corresponding to irreducible flat vector bundles is thus preserved under the transformation group $P\ \mathrm{Aut}(X)$; and $P\ \mathrm{Aut}(X)$ acts freely as a complex Lie group of nonsingular affine transformations on the subset $\Gamma_0(M, \mathcal{O}^{1,0}(\mathrm{Ad}\ X))$ of the complex vector space $\Gamma(M, \mathcal{O}^{1,0}(\mathrm{Ad}\ X))$.

Proof. Suppose that a connection $\lambda \in \Gamma(M, \mathcal{O}^{1,0}(\mathrm{Ad}\ X))$ is left fixed by an element $T \in P\ \mathrm{Aut}(X)$, $T \neq I$; thus $\lambda_\alpha = \mathrm{Ad}(T_\alpha)\cdot(\lambda_\alpha - DT_\alpha)$, where $(T_\alpha) \in \mathrm{Aut}(X)$ is an automorphism which is not given by a scalar matrix, that is, which is not of the form cI . Write $\lambda_\alpha = DF_\alpha$ for some functions $F_\alpha \in GL(n, \mathcal{O}_{U_\alpha})$, so that the cocycle $(X'_{\alpha\beta}) = (F_\alpha \cdot X_{\alpha\beta} \cdot F_\beta^{-1})$ represents the flat vector bundle corresponding to the connection λ ; and set $S_\alpha = F_\alpha \cdot T_\alpha \cdot F_\alpha^{-1}$ in each set U_α . Note that

$$\begin{aligned} DS_\alpha &= \mathrm{Ad}(F_\alpha T_\alpha^{-1})\cdot DF_\alpha + \mathrm{Ad}(F_\alpha)\cdot DT_\alpha - \mathrm{Ad}(F_\alpha)\cdot DF_\alpha \\ &= \mathrm{Ad}(F_\alpha T_\alpha^{-1})\cdot[\lambda_\alpha - \mathrm{Ad}(T_\alpha)\cdot(\lambda_\alpha - DT_\alpha)] = 0 , \end{aligned}$$

so that the matrices S_α are nonsingular constant matrices; and note further that

$$\begin{aligned} S_\alpha \cdot X'_{\alpha\beta} &= F_\alpha T_\alpha F_\alpha^{-1} \cdot F_\alpha X_{\alpha\beta} F_\beta^{-1} \\ &= F_\alpha X_{\alpha\beta} F_\beta^{-1} \cdot F_\beta T_\beta F_\beta^{-1} = X'_{\alpha\beta} \cdot S_\beta . \end{aligned} \tag{18}$$

The matrices S_α are not scalar matrices, since T_α are not scalar matrices; and it follows immediately from (18) then that the flat vector bundle represented by the cocycle $(X'_{\alpha\beta})$ is reducible.

Conversely, if the flat vector bundle represented by the cocycle $(\chi'_{\alpha\beta})$ associated to the connection λ is reducible, there exist nonsingular constant matrices S_α which are not scalar matrices, but which satisfy $S_\alpha \cdot \chi'_{\alpha\beta} = \chi'_{\alpha\beta} \cdot S_\beta$. Reversing the preceding argument, the matrices $T_\alpha = F_\alpha^{-1} \cdot S_\alpha \cdot F_\alpha$ represent a non-scalar automorphism of the bundle $(\chi_{\alpha\beta})$ such that $Ad^*(T) \cdot \lambda = \lambda$. This proves the first statement of the lemma; and since the second statement is an immediate consequence of the first, the proof is thereby concluded.

Since the complex Lie group $P\,Aut(X)$ acts freely as a group of complex analytic automorphisms of the complex manifold $\Gamma_o(M, \mathcal{O}^{1,0}(Ad\ X))$, the natural supposition is that the analogue of Theorem 27 holds, that is, that the quotient space

$$(21) \qquad \Lambda_o^*(M,X) = \Gamma_o(M, \mathcal{O}^{1,0}(Ad\ X))/P\,Aut(X)$$

has the structure of a complex analytic manifold such that the natural projection

$$(22) \qquad \pi\colon \Gamma_o(M, \mathcal{O}^{1,0}(Ad\ X)) \longrightarrow \Lambda_o^*(M,X)$$

is a complex analytic principal $P\,Aut(X)$ bundle. The individual orbits locally are submanifolds, and the manifold $\Gamma_o(M, \mathcal{O}^{1,0}(Ad\ X))$ locally has such a product structure, on general principles. In order to prove the supposition, referring back to the proof of Theorem 27, it is only necessary to establish an analogue of Lemma 28 for the action of the group $P\,Aut(X)$. We shall establish this result for bundles of rank $n = 2$ in the course of a more explicit analysis in the subsequent paragraphs. Before turning to this, however, it is interesting to see what can be said about the complex manifold $\Lambda_0^*(M,X)$ in general, assuming the truth of this supposition, and with as little work as possible.

First, it is easy to see that <u>the complex manifold $\Lambda_0^*(M,X)$ has complex dimension</u> $n^2(g-1)+1$ for any bundle $X \in H^1(M, GL(n,\mathbf{C}))$, where g is the genus of the compact Riemann surface M. (This is also assuming that the manifold $\Lambda_0^*(M,X)$ is not empty, of course.) For from the fibration (22) and earlier observations it follows that

$$\dim \Lambda_0^*(M,X) = \dim \Gamma_0(M, \mathcal{O}^{1,0}(\mathrm{Ad}\ X)) - \dim P\,\mathrm{Aut}(X)$$

$$= \dim \Gamma(M, \mathcal{O}^{1,0}(\mathrm{Ad}\ X)) - \dim \Gamma(M, \mathcal{O}(\mathrm{Ad}\ X)) + 1 ;$$

while from the Riemann-Roch theorem in the form given by formula (9) of §4, applied to the bundle $\mathrm{Ad}\ X$, it follows that

$$\dim \Gamma(M, \mathcal{O}(\mathrm{Ad}\ X)) - \dim \Gamma(M, \mathcal{O}^{1,0}(\mathrm{Ad}\ X^*)) = n^2(1-g) .$$

Since the bundles $\mathrm{Ad}\ X$ and $\mathrm{Ad}\ X^*$ are canonically isomorphic, the desired result follows immediately. By the way, if one is interested only in bundles of determinant one, hence in connections of trace zero, the corresponding space has dimension $n^2(g-1)$; in many ways, this is the more natural space to consider.

Next, <u>the tangent space to the manifold $\Lambda_0^*(M,X)$ at the point corresponding to the trivial connection $\lambda = 0 \in \Gamma(M, \mathcal{O}^{1,0}(\mathrm{Ad}\ X))$ can be identified in a natural manner with the vector space</u>

(23) $$\Gamma(M, \mathcal{O}^{1,0}(\mathrm{Ad}\ X))/d\Gamma(M, \mathcal{O}(\mathrm{Ad}\ X)) .$$

On the one hand, since $\Gamma_0(M, \mathcal{O}^{1,0}(\mathrm{Ad}\ X))$ is an open subset of the complex vector space $\Gamma(M, \mathcal{O}^{1,0}(\mathrm{Ad}\ X))$, the tangent space to $\Gamma_0(M, \mathcal{O}^{1,0}(\mathrm{Ad}\ X))$ at any point can be identified with the vector space $\Gamma(M, \mathcal{O}^{1,0}(\mathrm{Ad}\ X))$ itself. On the other hand, since $\mathrm{Aut}(X)$ is the group of invertible elements in the algebra $\mathrm{End}(X)$, the tangent space to $\mathrm{Aut}(X)$ at the identity can be identified with the space $\mathrm{End}(X) = \Gamma(M, \mathcal{O}(\mathrm{Ad}\ X))$; explicitly, if $T(t) = (T_\alpha(t))$ is a one-parameter subgroup of $\mathrm{Aut}(X)$, then

$$A_\alpha = \frac{d}{dt} T_\alpha(t)\Big|_{t=0} \in \Gamma(M, \mathcal{O}(\mathrm{Ad}\ X)) .$$

Now the orbit of the trivial connection $\lambda = 0 \in \Gamma(M, \mathcal{O}^{1,0}(\mathrm{Ad}\ X))$ under the action of a one-parameter subgroup $T(t) \subset \mathrm{Aut}(X)$ is given by $(\mathrm{Ad}^*(T(t))\cdot\lambda)_\alpha = -dT_\alpha(t)\cdot T_\alpha(t)^{-1}$; and since

$$\begin{aligned}\frac{d}{dt}(\mathrm{Ad}^*(T(t))\cdot\lambda)_\alpha\Big|_{t=0} &= \left[-d\left(\frac{dT_\alpha}{dt}\right) T_\alpha^{-1} + dT_\alpha T_\alpha^{-1} \frac{dT_\alpha}{dt} T_\alpha^{-1}\right]\Bigg|_{t=0} \\ &= -dA_\alpha \in d\Gamma(M, \mathcal{O}(\mathrm{Ad}\ X)) ,\end{aligned}$$

the tangents to the orbits of the group $\mathrm{Aut}(X)$ at the trivial connection $\lambda = 0$ form the vector subspace $d\Gamma(M, \mathcal{O}(\mathrm{Ad}\ X)) \subset \Gamma(M, \mathcal{O}^{1,0}(\mathrm{Ad}\ X))$, which suffices to conclude the proof of the result. In connection with this observation, it should be noted that the assertion is vacuous unless the bundle X is itself irreducible. If X is irreducible, the constant sections of the bundle $\mathrm{Ad}(X)$ form a one-dimensional family; thus $\dim d\Gamma(M, \mathcal{O}(\mathrm{Ad}\ X)) = \dim \Gamma(M, \mathcal{O}(\mathrm{Ad}\ X)) - 1$, and the dimension of the tangent space (23) agrees with the dimension of the manifold

$\Lambda_o^*(M,X)$ as calculated in the preceding paragraph.

(d) Considering in more detail the special case of bundles of rank $n = 2$, recall from §5 that the complex analytic vector bundle corresponding to a flat complex vector bundle $X \in H^1(M, GL(2,\mathbf{C}))$ can be represented by a cocycle of the form

$$(24) \qquad (\Phi_{\alpha\beta}) = \begin{pmatrix} \varphi_{1\alpha\beta} & \tau_{\alpha\beta} \\ 0 & \varphi_{2\alpha\beta} \end{pmatrix} \in Z^1(\mathfrak{U}, \mathcal{GL}(2,\mathcal{O})) ;$$

the components $(\varphi_{i\alpha\beta})$ are cocycles representing complex line bundles φ_i such that $\varphi_1 \subset X$ and $\varphi_2 \cong X/\varphi_1$, and it can be assumed that $c(\varphi_1) = \text{div } X$. (Recall that div X is the maximum value of the Chern classes of complex line bundles which can appear as subbundles of X.) Since X is a flat bundle, it follows that $c(\varphi_1) + c(\varphi_2) = 0$, and that X is analytically indecomposable unless $c(\varphi_1) = c(\varphi_2) = 0$; and furthermore, $-g \leqq c(\varphi_1) \leqq g-1$, where g is the genus of the Riemann surface M. If $c(\varphi_1) > 0$, or if $c(\varphi_1) = 0$ and the bundle X is analytically indecomposable, then the line bundle φ_1 is uniquely determined; indeed, φ_1 is the unique line bundle such that $\varphi_1 \subset X$ and $c(\varphi_1) \geqq 0$.

Paralleling the discussion on page 200, there is a complex analytic mapping

$$\Lambda(\mathfrak{U}, X) \longrightarrow \Lambda(\mathfrak{U}, (\Phi_{\alpha\beta}))$$

commuting with the actions (17) of the complex analytic transformation group $\text{Aut}(X)$; so the discussion of the structure of the et of equivalence classes of complex analytic connections for the undle X can be translated into terms of complex analytic connec-ions for the cocycle (24). This is the key to the following dis-ussion.

Remarks. In the discussion of complex analytic connections n §6(c), connections were actually only defined for a specific ocycle representing the complex analytic vector bundle in terms f the given open covering $\mathfrak{U}$; thus the set of connections pro-ably should have been denoted by $\Lambda(\mathfrak{U},(\Phi_{\alpha\beta}))$ rather than by $(\mathfrak{U},\Phi)$. Whenever cocycles $(\Phi_{\alpha\beta})$ and $(\Phi'_{\alpha\beta})$ represent the ame complex analytic vector bundle, formula (12) can be used to stablish an isomorphism between the sets $\Lambda(\mathfrak{U},(\Phi_{\alpha\beta}))$ and $(\mathfrak{U},(\Phi'_{\alpha\beta}))$; the temptation is to call the connections related y this isomorphism equivalent, and to define the set of connec-ions $\Lambda(\mathfrak{U},\Phi)$ for the vector bundle Φ as the set of equiva-ence classes of connections for all the cocycles representing the undle Φ . The mapping (12) is not always uniquely defined, how-ver; so that it is first necessary to pass to the set of equiva-ence classes $\Lambda(\mathfrak{U},(\Phi_{\alpha\beta}))/\mathrm{Aut}(\Phi) = \Lambda^*(\mathfrak{U},(\Phi_{\alpha\beta}))$. When $(\Phi_{\alpha\beta})$ s a flat cocycle, the set $\Lambda(\mathfrak{U},(\Phi_{\alpha\beta}))$ can be identified with he set $\Gamma(M,\mathcal{O}^{1,0}(\mathrm{Ad}\ \Phi))$ of Prym differentials for the flat vector undle, and the complications are less.

If $\lambda = (\lambda_\alpha) \in \Lambda(\mathfrak{U},(\Phi_{\alpha\beta}))$ is a complex analytic connec-ion for the cocycle (24), the terms λ_α and 2×2 matrices of olomorphic differential forms in the various open sets U_α of the covering $\mathfrak{U}$, such that in each intersection $U_\alpha \cap U_\beta$,

$$D\Phi_{\alpha\beta} = \delta(\lambda_\alpha) = \lambda_\beta - \mathrm{Ad}(\Phi_{\alpha\beta}^{-1})\cdot\lambda_\alpha \ .$$

This equation can be rewritten more explicitly as

(25)

$$\begin{pmatrix} \varphi_{1\alpha\beta} & \tau_{\alpha\beta} \\ 0 & \varphi_{2\alpha\beta} \end{pmatrix}\begin{pmatrix} \lambda_{11\beta} & \lambda_{12\beta} \\ \lambda_{21\beta} & \lambda_{22\beta} \end{pmatrix} - \begin{pmatrix} \lambda_{11\alpha} & \lambda_{12\alpha} \\ \lambda_{21\alpha} & \lambda_{22\alpha} \end{pmatrix}\begin{pmatrix} \varphi_{1\alpha\beta} & \tau_{\alpha\beta} \\ 0 & \varphi_{2\alpha\beta} \end{pmatrix} = \begin{pmatrix} d\varphi_{1\alpha\beta} & d\tau_{\alpha\beta} \\ 0 & d\varphi_{2\alpha\beta} \end{pmatrix} .$$

A more detailed analysis of the component functions of such a connection is not of great interest in general; but the following simple observations will be of use later.

<u>Lemma 31.</u> If there exists a complex analytic connection $\lambda = (\lambda_\alpha) \in \Lambda(\mathfrak{U},(\Phi_{\alpha\beta}))$ for the cocycle (24) such that $\lambda_{21\alpha} = 0$, then necessarily $c(\varphi_1) = c(\varphi_2) = 0$; and if $c(\varphi_1) = c(\varphi_2) = 0$, there exist complex analytic connections $\lambda = (\lambda_\alpha) \in \Lambda(\mathfrak{U},(\Phi_{\alpha\beta}))$ for the cocycle (24) such that $\lambda_{21\alpha} = 0$ and $\lambda_{11\alpha} \neq \lambda_{22\alpha}$, on surfaces of genus $g > 1$.

Proof. If $\lambda = (\lambda_\alpha)$ is a complex analytic connection for which $\lambda_{21\alpha} = 0$, then writing out equation (25) in detail, it follows that the remaining components of the connection λ_α are arbitrary analytic differential forms subject only to the conditions

$$(26) \begin{cases} \lambda_{11\beta} - \lambda_{11\alpha} = d \log \varphi_{1\alpha\beta} \\ \lambda_{22\beta} - \lambda_{22\alpha} = d \log \varphi_{2\alpha\beta} \\ \lambda_{12\beta} - \varphi_{1\alpha\beta}^{-1}\varphi_{2\alpha\beta}\lambda_{12\alpha} = (\lambda_{11\alpha} - \lambda_{22\beta})\varphi_{1\alpha\beta}^{-1}\tau_{\alpha\beta} + \varphi_{1\alpha\beta}^{-1}d\tau_{\alpha\beta} \end{cases}$$

The first line in (26) shows that the cohomology class in $H^1(M, \mathcal{O}^{1,0})$ represented by the cocycle $(d \log \varphi_{1\alpha\beta})$ is trivial, hence as in Lemma 19 of last year's lectures, necessarily $c(\varphi_1) = 0$; applying the same argument to the second line in (26), or recalling that $c(\varphi_1) + c(\varphi_2) = 0$, it also follows that $c(\varphi_2) = 0$. Now assume that $c(\varphi_1) = c(\varphi_2) = 0$; the bundles φ_1 and φ_2 have flat representatives, so that the cocycles $\varphi_{i\alpha\beta}$ can be taken to be constants, hence $d \log \varphi_{1\alpha\beta} = d \log \varphi_{2\alpha\beta} = 0$. The first two lines in (26) then merely assert that $\lambda_{11} = \lambda_{11\alpha} = \lambda_{11\beta}$ and $\lambda_{22} = \lambda_{22\alpha} = \lambda_{22\beta}$ are global holomorphic differential forms on the Riemann surface; and in order that there should exist holomorphic differential forms $\lambda_{12\alpha}$ satisfying the third line in (26), the difference $\sigma = \lambda_{11} - \lambda_{22}$ must be an abelian differential such that the cohomology class in $H^1(M, \mathcal{O}^{1,0}(\varphi_1^{-1}\varphi_2))$ represented by the cocycle

(27) $$\sigma\, \varphi_{1\alpha\beta}^{-1} \tau_{\alpha\beta} + \varphi_{1\alpha\beta}^{-1} d\tau_{\alpha\beta}$$

is trivial. By the Serre duality theorem, $H^1(M, \mathcal{O}^{1,0}(\varphi_1^{-1}\varphi_2)) \cong \Gamma(M, \mathcal{O}(\varphi_1\varphi_2^{-1}))$. If $\varphi_1 \neq \varphi_2$, then $\Gamma(M, \mathcal{O}(\varphi_1\varphi_2^{-1})) = 0$, so that σ can be completely arbitrary. If $\varphi_1 = \varphi_2$, then $\Gamma(M, \mathcal{O}(\varphi_1\varphi_2^{-1})) = \Gamma(M, \mathcal{O}) \cong \mathbf{C}$; the mapping which takes an abelian differential σ into the cohomology class represented by the cocycle (27) is a linear mapping $\Gamma(M, \mathcal{O}^{1,0}) \longrightarrow \mathbf{C}$, and since the surface M has genus $g > 1$, there will be nontrivial abelian differentials σ in the kernel of this mapping. That suffices to conclude the proof of the lemma.

Now consider an element $T \in P\,\mathrm{Aut}(X)$. If the bundle X is represented by the cocycle (24), then a representative $(T_\alpha) \in \mathrm{Aut}(X) = \mathrm{Aut}(\Phi_{\alpha\beta})$ of T will consist of a collection of matrices $T_\alpha \in GL(2, \mathcal{O}_{U_\alpha})$ such that $T_\alpha \cdot \Phi_{\alpha\beta} = \Phi_{\alpha\beta} \cdot T_\beta$ in $U_\alpha \cap U_\beta$. This can be written more explicitly as

$$(28) \quad \begin{pmatrix} t_{11\alpha} & t_{12\alpha} \\ t_{21\alpha} & t_{22\alpha} \end{pmatrix} \begin{pmatrix} \varphi_{1\alpha\beta} & \tau_{\alpha\beta} \\ 0 & \varphi_{2\alpha\beta} \end{pmatrix} = \begin{pmatrix} \varphi_{1\alpha\beta} & \tau_{\alpha\beta} \\ 0 & \varphi_{2\alpha\beta} \end{pmatrix} \begin{pmatrix} t_{11\beta} & t_{12\beta} \\ t_{21\beta} & t_{22\beta} \end{pmatrix} .$$

Using this, the group $P\,\mathrm{Aut}(X)$ can be described quite simply in terms of the analytic invariants of the bundle X, as follows.

Theorem 29. Let $X \in H^1(M, GL(2,\mathbf{C}))$ be a flat vector bundle of rank 2 on a compact Riemann surface M of genus $g > 1$.

(i) If X represents a stable complex analytic vector bundle, then $P\,\mathrm{Aut}(X) = 1$, the trivial group.

(ii) If X represents an indecomposable unstable complex analytic vector bundle, then $P\,\mathrm{Aut}(X) \cong \Gamma(M, \mathcal{O}(\varphi_1\varphi_2^{-1}))$.

(iii) If $X = \varphi_1 \oplus \varphi_2$ is a decomposable flat vector bundle, then

$$P\,\mathrm{Aut}(X) \cong \begin{cases} \mathbf{C}^* & \text{if } \varphi_1 \not\doteq \varphi_2 \\ PGL(2,\mathbf{C}) & \text{if } \varphi_1 \doteq \varphi_2 \end{cases} .$$

Proof. (i) For any non-trivial element $T \in P\,\mathrm{Aut}(X)$, it follows from Lemma 19 that there are representatives $(\Phi_{\alpha\beta}) \in Z^1(\mathfrak{U}, \mathfrak{gl}(2, \mathcal{O}))$ of X and $(T_\alpha) \in C^0(\mathfrak{U}, \mathfrak{gl}(2, \mathcal{O}))$ of T such that the matrices $\Phi_{\alpha\beta}$ and T_α are all in upper triangular form; the matrices $\Phi_{\alpha\beta}$ will still be written out with the notation used in (24), although of course it is not necessarily true that $c(\varphi_1) = \operatorname{div}(X)$. Writing out equation (28) in detail, and remembering that $t_{21\alpha} = t_{21\beta} = 0$, it follows that

$$(29) \qquad \begin{cases} t_{11\alpha} = t_{11\beta}, \qquad t_{22\alpha} = t_{22\beta}, \\ t_{12\beta} - \varphi_{1\alpha\beta}^{-1}\varphi_{2\alpha\beta}t_{12\alpha} = (t_{11\alpha} - t_{22\beta})\varphi_{1\alpha\beta}^{-1}\tau_{\alpha\beta} \end{cases} .$$

The first line in (29) shows that $t_{11} = t_{11\alpha} = t_{11\beta}$ and $t_{22} = t_{22\alpha} = t_{22\beta}$ are holomorphic functions on the entire Riemann surface, hence constants. If $t_{11} \neq t_{22}$, the second line in (29) shows that the cohomology class in $H^1(M, \mathcal{O}(\varphi_1^{-1}\varphi_2))$ represented by the cocycle $(\varphi_{1\alpha\beta}^{-1}\tau_{\alpha\beta})$ is trivial; it then follows from Theorem 13 that the bundle X is analytically decomposable, hence is unstable. If $t_{11} = t_{22}$, the second line in (29) shows that $t_{12\alpha} = \varphi_{1\alpha\beta}\varphi_{2\alpha\beta}^{-1}t_{12\beta}$. Since $T \in P\operatorname{Aut}(X)$ is non-trivial, $t_{12\alpha} \neq 0$; hence $2c(\varphi_1) = c(\varphi_1\varphi_2^{-1}) \geqq 0$, and the bundle X is again unstable. Altogether then, if $P\operatorname{Aut}(X)$ is a non-trivial group, the complex analytic vector bundle X is necessarily unstable; and this proves part (i) of the theorem.

(ii) If X is an unstable complex analytic vector bundle, select a representative cocycle $(\Phi_{\alpha\beta}) \in Z^1(\mathfrak{U}, \mathfrak{b}\mathfrak{L}(2, \mathcal{O}))$ in the standard form (24); and for any element $T \in P\operatorname{Aut}(X)$, select a representative (T_α) explicitly as in (28). Considering the components in the second row and first column of (28), note that $t_{21\alpha} = \varphi_{1\alpha\beta}^{-1}\varphi_{2\alpha\beta} \cdot t_{21\beta}$. Since the bundle X is unstable,

$c(\varphi_1^{-1}\varphi_2) = -2c(\varphi_1) \leqq 0$; therefore $t_{21\alpha} = 0$, except in the case that $\varphi_1 = \varphi_2$, when $t_{21} = t_{21\alpha} = t_{21\beta}$ is an arbitrary complex constant. In this latter case though, considering the components in the first row and first column of (28), it follows that

$$t_{11\beta} - t_{11\alpha} = t_{21} \cdot \varphi_{1\alpha\beta}^{-1} \tau_{\alpha\beta} ;$$

and as before, if $t_{21} \neq 0$ the bundle X is analytically decomposable. In any case then, necessarily $t_{21\alpha} = 0$; so the conditions (28) again reduce to the conditions (29). As in part (i) above, the first line in (29) shows that $t_{11} = t_{11\alpha} = t_{11\beta}$ and $t_{22} = t_{22\alpha} = t_{22\beta}$ are complex constants; and since the bundle X is analytically indecomposable, the second line in (29) shows that $t_{11} = t_{22}$ and that $(t_{12\alpha}) \in \Gamma(M, \mathcal{O}(\varphi_1\varphi_2^{-1}))$. Multiplying the matrices T_α by a complex constant, the element $T \in P\,\mathrm{Aut}(X)$ has a representative (T_α) in the form

(30) $$T_\alpha = \begin{pmatrix} 1 & t_{12\alpha} \\ 0 & 1 \end{pmatrix}, \text{ for any } t_{12\alpha} \in \Gamma(M, \mathcal{O}(\varphi_1\varphi_2^{-1})) ;$$

and it is obvious that this establishes a one-to-one correspondence, indeed a Lie group isomorphism, between the groups $P\,\mathrm{Aut}(X)$ and $\Gamma(M, \mathcal{O}(\varphi_1\varphi_2^{-1}))$.

(iii) If X is an analytically decomposable vector bundle, select a representative cocycle $(\Phi_{\alpha\beta}) \in Z^1(\mathfrak{U}, \mathfrak{g}\mathfrak{l}(2, \mathcal{O}))$ in the standard form (24), with $\varphi_{1\alpha\beta}$ complex constants, and $\tau_{\alpha\beta} = 0$; and consider an element $T \in P\,\mathrm{Aut}(X)$, with a representative (T_α) as in (28). If $\varphi_1 \neq \varphi_2$, it follows as in part (ii) that $t_{21\alpha} = 0$, and that $t_{11} = t_{11\alpha} = t_{11\beta}$ and $t_{22} = t_{22\alpha} = t_{22\beta}$ are constants; and since $(t_{12\alpha}) \in \Gamma(M, \mathcal{O}(\varphi_1\varphi_2^{-1})) = 0$, it further follows that $t_{12\alpha} = 0$. Multiplying the matrices T_α by a complex constant, the element $T \in P\,\mathrm{Aut}(X)$ has a representative (T_α) of the form

$$(31) \qquad T_\alpha = \begin{pmatrix} 1 & 0 \\ 0 & t_{22} \end{pmatrix}, \text{ for any } t_{22} \in \mathbb{C}^* ;$$

and this establishes the isomorphism $P\,\mathrm{Aut}(X) \cong \mathbb{C}^*$. Finally, if $\varphi_1 = \varphi_2$, equation (28) reduces to $T_\alpha = T_\beta$; so the matrices $T = T_\alpha = T_\beta \in GL(2,\mathbb{C})$ are arbitrary, hence clearly $P\,\mathrm{Aut}(X) \cong PGL(2,\mathbb{C})$. The proof of the theorem is thereby concluded.

Finally, consider the action of the complex Lie group $P\,\mathrm{Aut}(X)$ as a group of complex analytic automorphisms of the complex manifold $\Lambda(\mathfrak{U}, X) \cong \Lambda(\mathfrak{U}, (\Phi_{\alpha\beta}))$. If $T \in P\,\mathrm{Aut}(X)$ is represented by a transformation $(T_\alpha) \in \mathrm{Aut}(\Phi_{\alpha\beta})$, and if $\lambda = (\lambda_\alpha) \in \Lambda(\mathfrak{U}, (\Phi_{\alpha\beta}))$ is a complex analytic connection, the group action as defined by (17) is

$$(\mathrm{Ad}^*(T)\cdot\lambda)_\alpha = T_\alpha \lambda_\alpha T_\alpha^{-1} - dT_\alpha T_\alpha^{-1} .$$

Using the detailed description of the group $P\,\mathrm{Aut}(X)$ as provided by Theorem 29, the supposition of page 208 can be verified in this case as follows.

<u>Theorem 30.</u> If $X \in H^1(M, GL(2,\mathbb{C}))$ is a flat vector bundle of rank 2 on a compact Riemann surface M of genus $g > 1$, then the quotient space

$$\Lambda_0^*(M,X) = \Gamma_0(M, \mathcal{O}^{1,0}(\mathrm{Ad}\ X))/P\,\mathrm{Aut}(X)$$

has the structure of a complex analytic manifold of complex dimension $4g-3$ such that the natural projection

$$\pi\colon \Gamma_0(M, \mathcal{O}^{1,0}(\mathrm{Ad}\ X)) \longrightarrow \Lambda_0^*(M, X)$$

is a complex analytic principal $P\,\mathrm{Aut}(X)$ bundle.

Proof. If X represents a stable complex analytic vector bundle, then from part (i) of Theorem 29 the group $P\,\mathrm{Aut}(X)$ is trivial; the desired result is an immediate consequence. Note that all flat vector bundles analytically equivalent to X are necessarily irreducible, so that $\Gamma_0(M, \mathcal{O}^{1,0}(\mathrm{Ad}\ X)) = \Gamma(M, \mathcal{O}^{1,0}(\mathrm{Ad}\ X))$ is non-empty; the formula for the dimension of the space $\Lambda_0^*(M, X)$ was derived on page 209.

If X represents an unstable indecomposable complex analytic vector bundle, then from part (ii) of Theorem 29 there is an isomorphism $P\,\mathrm{Aut}(X) \cong \Gamma(M, \mathcal{O}(\varphi_1\varphi_2^{-1}))$; the isomorphism is given explicitly by (30). Writing out the group action (17) in detail, for the element $T \in P\,\mathrm{Aut}(X)$ corresponding to a section $(t_\alpha) \in \Gamma(M, \mathcal{O}(\varphi_1\varphi_2^{-1}))$ and for any connection $\lambda = (\lambda_\alpha) \in \Lambda(\mathfrak{U}, (\Phi_{\alpha\beta}))$,

$$(\mathrm{Ad}^*(T)\cdot\lambda)_\alpha = \begin{pmatrix} 1 & t_\alpha \\ 0 & 1 \end{pmatrix}\begin{pmatrix} \lambda_{11\alpha} & \lambda_{12\alpha} \\ \lambda_{21\alpha} & \lambda_{22\alpha} \end{pmatrix}\begin{pmatrix} 1 & -t_\alpha \\ 0 & 1 \end{pmatrix} - \begin{pmatrix} 0 & dt_\alpha \\ 0 & 0 \end{pmatrix}\begin{pmatrix} 1 & -t_\alpha \\ 0 & 1 \end{pmatrix}$$

(32)

$$= \begin{pmatrix} \lambda_{11\alpha} + t_\alpha\lambda_{21\alpha} & \lambda_{12\alpha} + t_\alpha(\lambda_{22\alpha} - \lambda_{11\alpha}) - t_\alpha^2\lambda_{21\alpha} - dt_\alpha \\ \lambda_{21\alpha} & \lambda_{22\alpha} - t_\alpha\lambda_{21\alpha} \end{pmatrix}.$$

First, consider the case that $c(\varphi_1) = \mathrm{div}(X) > 0$; all flat vector bundles analytically equivalent to X are irreducible, so that $\Gamma_0(M, \mathcal{O}^{1,0}(\mathrm{Ad}\ X)) = \Gamma(M, \mathcal{O}^{1,0}(\mathrm{Ad}\ X))$ is non-empty. (To see the same thing in a different manner, by Lemma 30 the connections corresponding to reducible flat vector bundles are fixed points for some transformations of the group $P\,\mathrm{Aut}(X)$. If $\lambda = (\lambda_\alpha)$ is fixed for a non-trivial element $T \in P\,\mathrm{Aut}(X)$, it is clear from (32) that $\lambda_{21\alpha} = 0$; but since $c(\varphi_1) > 0$, Lemma 31 shows that this is impossible.) The Lie group $P\,\mathrm{Aut}(X)$ acts freely as a group of complex analytic automorphisms of the complex manifold $\Lambda(\mathcal{U}, (\Phi_{\alpha\beta}))$. Upon examining the proof of Theorem 27, it is clear that it suffices merely to prove an analogue of Lemma 28 in this case. The proof of this analogue is quite trivial, though; for since $\lambda_{21\alpha} \neq 0$ by Lemma 31, it follows from (32) that the group $P\,\mathrm{Aut}(X) \cong \Gamma(M, \mathcal{O}(\varphi_1\varphi_2^{-1}))$ acts as a non-trivial group of translations on the components in the first row and first column of the connections, and the desired result is an immediate consequence. Next, considering the case that $c(\varphi_1) = c(\varphi_2) = \mathrm{div}(X) = 0$,

$$P\,\mathrm{Aut}(X) \cong \Gamma(M, \mathcal{O}(\varphi_1\varphi_2^{-1})) = \begin{cases} 0 & \text{if } \varphi_1 \neq \varphi_2, \\ \mathbf{C} & \text{if } \varphi_1 = \varphi_2; \end{cases}$$

the desired result being trivial if $\varphi_1 \neq \varphi_2$, there only remains the subcase $\varphi_1 = \varphi_2$. The action of the group $P\,\mathrm{Aut}(X) \cong \mathbf{C}$ on the manifold $\Lambda(\mathcal{U}, (\Phi_{\alpha\beta}))$ is given by (32), where $t = t_\alpha \in \mathbf{C}$ is an arbitrary complex constant. Note that the fixed points of a transformation $t \neq 0$ are those complex analytic connections $\lambda = (\lambda_\alpha)$ such that $\lambda_{21\alpha} = 0$ and $\lambda_{22\alpha} = \lambda_{11\alpha}$; Lemma 31 shows that there are connections which are not left fixed by any non-trivial transformation, so that there are always irreducible flat vector bundles equivalent to the given bundle X. Restricting attention to these connections, it again follows immediately from (32) that the group $P\,\mathrm{Aut}(X)$ always acts as a non-trivial group of translations on some component of the connections, and the proof is completed as above.

Finally, if X represents a decomposable complex analytic vector bundle $X = \varphi_1 \oplus \varphi_2$, then upon writing (25) out explicitly for the case that $\tau_{\alpha\beta} = 0$, the set $\Lambda(\mathfrak{U},(\Phi_{\alpha\beta}))$ of complex analytic connections has the form

$$\lambda = (\lambda_\alpha) = (\lambda_{ij\alpha}) , \text{ for arbitrary } \lambda_{ij\alpha} \in \Gamma(M, \mathcal{O}^{1,0}(\varphi_i\varphi_j^{-1})) .$$

Note that $\dim \Gamma(M, \mathcal{O}^{1,0}(\varphi_i\varphi_j^{-1})) = g$ or $g-1$ according as $\varphi_i = \varphi_j$ or $\varphi_i \neq \varphi_j$; hence $\dim \Gamma(M, \mathcal{O}^{1,0}(\varphi_i\varphi_j^{-1})) > 0$ always. First, suppose that $\varphi_1 \neq \varphi_2$; so by part (iii) of Theorem 30, note that $P\,\mathrm{Aut}(X) \cong \mathbf{C}^*$, the isomorphism being given explicitly by (31). Writing out the group action (17) in detail, for the element $T \in P\,\mathrm{Aut}(X)$ corresponding to $c \in \mathbf{C}^*$ and for any connection $\lambda = (\lambda_\alpha) \in \Lambda(\;\; ,(\Phi_{\alpha\beta}))$,

$$(\mathrm{Ad}^*(T)\cdot\lambda)_\alpha = \begin{pmatrix} 1 & 0 \\ 0 & c \end{pmatrix}\begin{pmatrix} \lambda_{11\alpha} & \lambda_{12\alpha} \\ \lambda_{21\alpha} & \lambda_{22\alpha} \end{pmatrix}\begin{pmatrix} 1 & 0 \\ 0 & 1/c \end{pmatrix}$$

$$= \begin{pmatrix} \lambda_{11\alpha} & \frac{1}{c}\lambda_{12\alpha} \\ c\lambda_{21\alpha} & \lambda_{22\alpha} \end{pmatrix} .$$

The set of connections which are not fixed points of any non-trivial $T \in X\,\mathrm{Aut}(X)$ is described by $\lambda_{21\alpha}\cdot\lambda_{12\alpha} \neq 0$, and is obviously a non-empty set; the action of the group $\mathbf{C}^*$ obviously satisfies the analogue of Lemma 28, which completes the proof of this case. Lastly, suppose that $\varphi_1 = \varphi_2$, so that from part (iii) of Theorem 30 again, $P\,\mathrm{Aut}(X) \cong PGL(2,\mathbf{C})$. The connections are arbitrary matrices of abelian differentials, and the group $PGL(2,\mathbf{C})$ acts on these matrices by inner automorphism. The set of connections which are not fixed points of any non-trivial $P \in P\,\mathrm{Aut}(X) = PGL(2,\mathbf{C})$ is just the set of irreducible matrices of abelian differentials, and is clearly non-empty since $g > 1$. This case of the theorem reduces almost immediately to Theorem 27 itself; and with that, the proof of the entire theorem is concluded.

Remarks. In the course of the above proof it was demonstrated that the sets $\Lambda_0^*(M,X)$ are always non-empty. This can be restated as the assertion that every flat vector bundle $X \in H^1(M,GL(2,\mathbf{C}))$ is analytically equivalent to an irreducible flat vector bundle. It follows from this that in examining the set of complex analytic equivalence classes of flat vector bundles $X \in H^1(M,GL(2,\mathbf{C}))$, there is really no loss of generality at all in restricting attention to the set of irreducible flat vector bundles; thus the fact that the discussion in §9(b) was restricted to the subset $R_0 \subset R$, is of no great concern after all.

The manifolds $\Lambda_0^*(M,X)$ all have the same complex dimension ($4g-3$ for bundles of rank $n = 2$), as noted on page 209; but they are not all analytically homeomorphic. If X is a stable complex analytic vector bundle, or more generally if X is a bundle such that $P\,\mathrm{Aut}(X) = 1$ (a class of bundles called simple complex vector bundles, for the obvious reason), then $\Lambda_0^*(M,X) \cong \Gamma(M, \mathcal{O}^{1,0}(\mathrm{Ad}\ X)) \cong \mathbf{C}^{4g-3}$. If X represents a decomposable complex analytic vector bundle $X = \varphi_1 \oplus \varphi_2$ where $\varphi_1 \neq \varphi_2$, then as noted in the proof of Theorem 30,

$$\Lambda_0^*(M,X) \cong (\mathbf{C}^g \times \mathbf{C}^g \times E_{g-1} \times E_{g-1})/\mathbf{C}^* , \tag{33}$$

where $E_{g-1} = \mathbf{C}^{g-1} - (0,\dots,0)$ is the set of non-zero elements in $\mathbf{C}^{g-1}$ and the action of the group $\mathbf{C}^*$ is given by

$$c\cdot(z_1,z_2,w_1,w_2) = (z_1,z_2,cw_1,\tfrac{1}{c}w_2) ,$$

for $c \in \mathbf{C}^*$, $z_i \in \mathbf{C}^g$, $w_i \in E_{g-1}$. It is evident that the space (33) is not even topologically homeomorphic to the space $\mathbf{C}^{4g-3}$. It is quite possible to read off from the proof of Theorem 30 a description of the complex manifolds $\Lambda_0^*(M,X)$; the group actions are also interesting, remembering for instance the quadratic term in (32). But there is not time enough here to pursue this matter further.

The discussion on pages 197-201 of last year's lectures should be looked at in the light of the above discussion; see the appendix in R. C. Gunning, "Special coordinate coverings of Riemann surfaces," (Math. Annalen 170(1967), 67-86), for more details.

(e) Summarizing briefly the conclusions of the preceding parts (c) and (d), it is apparent that the mapping (11) induces a one-to-one complex analytic mapping between the following two complex manifolds

$$(34) \qquad H_X^0 \colon \Lambda_0^*(M,X) = \frac{\Gamma_0(M, \mathcal{O}^{1,0}(\mathrm{Ad}\ X))}{P\ \mathrm{Aut}(X)} \longrightarrow S_0 \cong H_0^1(M, GL(n,\mathbf{C})) \ ,$$

at least in the case of flat complex vector bundles X of rank $n = 2$; and the image of H_X^0 is the set of irreducible flat vector bundles analytically equivalent to X . Now it is a quite straightforward matter to describe the differential of this mapping; and the following observation then results.

Theorem 31. If $X \in H^1(M, GL(2,\mathbf{C}))$ is a flat complex vector bundle of rank 2 on a compact Riemann surface M of genus $g > 1$, then the mapping (34) is a regular mapping (has a non-singular differential); thus the image of H_X^0 is locally a complex analytic submanifold of the complex manifold S_0 , (in the sense that the image of a relatively compact subset of $\Lambda_0^*(M,X)$ is a submanifold). Identifying the tangent space of S_0 at X with the space $H^1(M, \mathcal{F}(\mathrm{Ad}\ X))$, the tangent space to the submanifold image (H_X^0) is the subspace of $H^1(M, \mathcal{F}(\mathrm{Ad}\ X))$ consisting of the period classes of the Prym differentials $\Gamma(M, \mathcal{O}^{1,0}(\mathrm{Ad}\ X))$.

Proof. It suffices merely to consider an open neighborhood of the trivial connection $\lambda = 0$, in view of the remarks on pages 200 and 205 . The tangent space to $\Gamma_0(M, \mathcal{O}^{1,0}(\mathrm{Ad}\ X))$ at $\lambda = 0$ is identified with the vector space $\Gamma(M, \mathcal{O}^{1,0}(\mathrm{Ad}\ X))$, as before. For any connection $\lambda = (\lambda_\alpha) \in \Gamma(M, \mathcal{O}^{1,0}(\mathrm{Ad}\ \chi))$, consider the one-parameter family of connections $t\lambda = (t\lambda_\alpha)$ for $t \in \mathbf{C}$; the image $H_\chi(t\lambda)$ of this family under the mapping (11) is a differentiable curve in the manifold S_0 , and the tangent vector to this curve at the point $H_\chi(0) = \chi$ is just the image vector $dH_\chi(\lambda)$. This latter tangent vector is of course just the derivative of the vector-valued function $H_\chi(t\lambda)$ at the point $t = 0$, when that function is expressed in local coordinates near χ .

To carry out this calculation explicitly, select a suitable open coordinate covering $\mathfrak{U} = \{U_\alpha\}$ of the Riemann surface M, with local coordinate z_α in U_α ; and select a base point $p_\alpha \in U_\alpha$ for each neighborhood. As in the proof of Lemma 29, for a suitable open neighborhood Δ of the origin in the complex t-plane, choose holomorphic functions $F_\alpha(z_\alpha, t) \in GL(2, \mathcal{O}_{U_\alpha \times \Delta})$ such that

$$(35) \qquad \begin{cases} \dfrac{\partial}{\partial z_\alpha} F_\alpha(z_\alpha, t)\, dz_\alpha = F_\alpha(z_\alpha, t)\cdot t\, \lambda_\alpha(z_\alpha) , \quad \text{and} \\[2ex] F_\alpha(p_\alpha, t) \equiv I \quad \text{for all} \quad t \in \Delta . \end{cases}$$

The cocycle

$$(36) \qquad (\chi_{\alpha\beta}(t)) = (F_\alpha(z_\alpha, t)\cdot\chi_{\alpha\beta}\cdot F_\beta(z_\beta, t)^{-1}) \in Z^1(\mathfrak{U}, GL(2, \mathbf{C}))$$

then represents the image $H_\chi(t\lambda)$ for each $t \in \Delta$; note that $F_\alpha(z_\alpha, 0) \equiv I$, hence that $\chi_{\alpha\beta}(0) = \chi_{\alpha\beta}$ is the given cocycle representing the flat vector bundle χ . Furthermore, choose holomorphic functions $G_\alpha(z_\alpha) \in \mathcal{O}^{2\times 2}_{U_\alpha}$ such that

$$(37) \qquad dG_\alpha(z_\alpha) = \lambda_\alpha(z_\alpha) \quad \text{and} \quad G_\alpha(p_\alpha) = 0 .$$

As on page 161, the constants

$$(38) \qquad A_{\alpha\beta} = G_{\beta}(z_{\beta}) - \mathrm{Ad}(X_{\alpha\beta}^{-1})\cdot G_{\alpha}(z_{\alpha})$$

form a one-cocycle $(A_{\alpha\beta}) \in Z^1(\mathfrak{U}, \mathcal{F}(\mathrm{Ad}\ X))$, which represents the period class $A = \delta(\lambda)$ of the Prym differential $\lambda \in \Gamma(M, \mathcal{O}^{1,0}(\mathrm{Ad}\ X))$. The cocycles (36) and (38) are related in the following manner. As noted already, it follows immediately from (35) that $F_{\alpha}(z_{\alpha}, 0) \equiv I$. Differentiating (35) with respect to t and setting $t = 0$, note that

$$\frac{\partial}{\partial z_{\alpha}}\left(\frac{\partial F_{\alpha}}{\partial t}(z_{\alpha}, 0)\right) dz_{\alpha} = \lambda_{\alpha}(z_{\alpha}) \ , \quad \frac{\partial F_{\alpha}}{\partial t}(p_{\alpha}, 0) \equiv 0 \ ;$$

consequently it is clear that $\frac{\partial F_{\alpha}}{\partial t}(z_{\alpha}, 0) \equiv G_{\alpha}(z_{\alpha})$. Then

$$(39) \qquad \begin{cases} X_{\alpha\beta}(t)^{-1}\cdot \frac{d}{dt} X_{\alpha\beta}(t)\Big|_{t=0} = X_{\alpha\beta}^{-1}[G_{\alpha}(z_{\alpha})X_{\alpha\beta} - X_{\alpha\beta}G_{\beta}(z_{\beta})] \\ \qquad\qquad\qquad\qquad\qquad\quad = -A_{\alpha\beta} \ . \end{cases}$$

To express the mapping $H_{X}(t\lambda)$ in terms of the complex coordinates introduced on the manifold $S_o \cong H_o^1(M, GL(2,\mathbf{C}))$ in parts (a) and (b) above, it is necessary to go from the cocycle (36) to the corresponding characteristic representation of the bundle. Let $\sigma = (U_{\alpha_o}, U_{\alpha_1}, \ldots, U_{\alpha_q})$ be a closed chain in $\pi_1(\mathfrak{U}, U_o)$ representing one of the standard generators of the fundamental group of the surface. The matrix associated to σ describing the characteristic representation corresponding to the cocycle (36) is

$$\hat{X}_{\sigma}(t) = X_{\alpha_o\alpha_1}(t)\cdot \ldots \cdot X_{\alpha_{q-1}\alpha_q}(t) \ ;$$

and the matrix associated to σ describing the cocycle in $Z^1(\pi_1(M),\hat{\chi})$ corresponding to the cocycle (38) under the isomorphism of Theorem 19 is

$$\hat{A}_\sigma = \sum_{j=1}^{q} Ad(X_{\alpha_j\alpha_{j+1}} \cdots X_{\alpha_{q-1}\alpha_q})^{-1} \cdot A_{\alpha_{j-1}\alpha_j} . \tag{4}$$

Introducing coordinates $X \in \mathbb{C}^{n\times n}$ in a neighborhood of the matrix $\hat{X}_\sigma(0)$ by the mapping

$$X \longrightarrow \hat{X}_\sigma(0)\cdot \exp X$$

as on page 182, the curve $H_\chi(t\lambda)$ is described in these coordinates by the curve $X(t)$ such that

$$\hat{X}_\sigma(0)\cdot \exp X(t) = \hat{X}_\sigma(t) .$$

Differentiating this last equation with respect to t and setting $t = 0$, it follows that

$$\frac{d}{dt} X(t)\Big|_{t=0} = \hat{X}_\sigma(0)^{-1} \cdot \sum_{j=1}^{q} X_{\alpha_0\alpha_1} \cdots X_{\alpha_{j-2}\alpha_{j-1}} \frac{dX_{\alpha_{j-1}\alpha_j}}{dt}(0) X_{\alpha_j\alpha_{j+1}} \cdots X_{\alpha_{q-1}\alpha_q}$$

$$= \sum_{j=1}^{q} Ad(X_{\alpha_j\alpha_{j+1}} \cdots X_{\alpha_{q-1}\alpha_q})^{-1} \cdot X^{-1}_{\alpha_{j-1}\alpha_j} \frac{dX_{\alpha_{j-1}\alpha_j}}{dt}(0)$$

$$= -\hat{A}_\sigma ,$$

by (39) and (40). It is thus clear that

$$dH_\chi(\lambda) = -\delta(\lambda) , \tag{41}$$

where $\delta(\lambda)$ denotes the period class of the Prym differential λ, and the tangent space to the manifold S_0 at the point X is identified with the cohomology group $H^1(\pi_1(M), Ad(X))$ as in Theorem 28.

Now to apply formula (41), recall from the remark on page 209 that the tangent space to the manifold $\Lambda_o^*(M,X)$ is naturally identified with the vector space $\Gamma(M, \mathcal{O}^{1,0}(\text{Ad } X))/d\Gamma(M, \mathcal{O}(\text{Ad } X))$. The differential of the complex analytic mapping

$$H_X^o\colon \Lambda_o^*(M,X) \longrightarrow S_o ,$$

then coincides with the negative of the period mapping

$$\delta\colon \frac{\Gamma(M, \mathcal{O}^{1,0}(\text{Ad } X))}{d\Gamma(M, \mathcal{O}(\text{Ad } X))} \longrightarrow H^1(M, \mathcal{F}(\text{Ad } X))$$

of the Prym differentials. It follows from Theorem 22 that this mapping is always an isomorphism into, and the proof of the theorem is therewith concluded.

This theorem shows that the complex analytic manifold S_o (of complex dimension $2(g-1)n^2+2$ where $n = 2$) is the disjoint union of the complex analytic submanifolds (each of complex dimension $(g-1)n^2+1$ where $n = 2$) consisting of the analytic equivalence classes of flat vector bundles; it must be recalled that these submanifolds have not been shown to be closed subsets, so they must provisionally be viewed as submanifolds in an extended sense. This splitting can also be described as the decomposition of the manifold S_o into integral submanifolds for the differential system

$$\delta\Gamma(M, \mathcal{O}^{1,0}(\text{Ad } X)) \subset H^1(M, \mathcal{F}(\text{Ad } X)) ,$$

where $H^1(M, \mathcal{F}(\mathrm{Ad}\ X))$ is identified with the tangent space to the manifold S_0 at the point $X \in S_0$. Upon identifying each of these submanifolds to a point, the resulting quotient space can be identified with the subset

$$H^1_*(M, \mathfrak{GL}(2,\mathcal{O})) \subset H^1(M, \mathfrak{GL}(2,\mathcal{O}))$$

consisting of those complex analytic vector bundles over M which admit flat representatives, a subset which was described quite explicitly in Weil's theorem, (Theorem 16); recall that it has been shown that all the vector bundles in this subset have irreducible flat representatives. The quotient mapping

$$\mu\colon S_0 \longrightarrow H^1_*(M, \mathfrak{GL}(2,\mathcal{O}))$$

can be viewed as a form of complex analytic fibration of the complex manifold S_0, but it must be a singular fibration in some sense since not all the fibres are even topologically the same; and this fibration induces some sort of complex analytic structure on the quotient space $H^1_*(M, \mathfrak{GL}(2,\mathcal{O}))$.

It is just at this stage, when things at last begin to look rather interesting and there are a considerable number of questions begging to be looked into, that time has unfortunately run out, and these lectures must be called to a halt. I hope to have an opportunity to continue the discussion of this subject in the near future. I cannot close without mentioning another approach to the imposition of a complex analytic structure, on the subspace of $H^1_*(M, \mathfrak{GL}(2,\mathcal{O}))$ consisting of those complex analytic vector bundles admitting unitary flat representatives, which can be found in the papers by M. S. Narasimhan and C. S. Seshadri ("Holomorphic vector bundles on a compact Riemann surface," Math. Annalen 155(1964), 69-80) and by C. S. Seshadri ("Space of unitary vector bundles on a compact Riemann surface," Annals of Math. 85 (1967), 303-336). An excellent survey of the literature and of the present general state of knowledge of complex vector bundles over arbitrary Riemann surfaces can be found in the paper by H. Röhrl ("Holomorphic fibre bundles over Riemann surfaces," Bull. Amer. Math. Soc. 68(1962), 125-160); the readers can find references there for the many topics not treated in these lectures.

Appendix 1. The formalism of cohomology with coefficients in a locally free analytic sheaf.

Several times in the present discussion an explicit description of the cohomology groups $H^q(M, \mathcal{S})$ with coefficients in a locally free analytic sheaf $\mathcal{S}$ over a Riemann surface M, has been required; the description involves the local isomorphism $\mathcal{S}|U \cong \mathcal{O}^m|U$, and is sometimes a bit confusing notationally, so an attempt will be made here to straighten things out.

Each locally free sheaf $\mathcal{S}$ of rank m is of course given by $\mathcal{S} = \mathcal{O}(\Phi)$ for some complex analytic vector bundle Φ. Select an open covering $\mathcal{U} = \{U_\alpha\}$ of the Riemann surface M, such that the sheaf $\mathcal{S}$ is free over each set U_α; and further, select an isomorphism $\mathcal{S}|U_\alpha \cong \mathcal{O}^m|U_\alpha$. Having made these choices, comparison of the isomorphisms over the intersections $U_\alpha \cap U_\beta$ yields a cocycle $(\Phi_{\alpha\beta}) \in Z^1(\mathcal{U}, \mathcal{GL}(m, \mathcal{O}))$ describing the vector bundle $\Phi \in H^1(M, \mathcal{GL}(m, \mathcal{O}))$ such that $\mathcal{S} = \mathcal{O}(\Phi)$, as in §2. The point now is that there is a useful description of the cohomology groups $H^q(\mathcal{U}, \mathcal{S}) = H^q(\mathcal{U}, \mathcal{O}(\Phi))$, as follows.

Recall that a cocycle $f \in Z^q(\mathcal{U}, \mathcal{S}) = Z^q(\mathcal{U}, \mathcal{O}(\Phi))$ is given by a collection of sections

(1) $$f_{\alpha_o\ldots\alpha_q} \in \Gamma(U_{\alpha_o} \cap\ldots\cap U_{\alpha_q}, \mathcal{S})$$

such that

(2) $$\sum_{j=0}^{q+1} (-1)^j f_{\alpha_o\ldots\alpha_{j-1}\alpha_{j+1}\ldots\alpha_{q+1}}(p) = 0 \quad \text{whenever} \quad p \in U_{\alpha_o} \cap\ldots\cap U_{\alpha_{q+1}}.$$

Since $U_{\alpha_o} \cap\ldots\cap U_{\alpha_q} \subset U_{\alpha_q}$, under the selected isomorphism

$\mathcal{S}|U_{\alpha_q} \cong \mathcal{O}^m|U_{\alpha_q}$ the section $f_{\alpha_0\ldots\alpha_q}$ can be identified with an element of the module $\Gamma(U_{\alpha_0}\cap\ldots\cap U_{\alpha_q}, \mathcal{O}^m)$; this section will be denoted by $f_{\alpha_0\ldots\alpha_q}(z_{\alpha_q})$, and is just a complex analytic mapping from the set $z_q(U_{\alpha_0}\cap\ldots\cap U_{\alpha_q}) \subset \mathbf{C}$ into $\mathbf{C}^m$, where z_{α_q} is the coordinate mapping in U_{α_q} . Of course, there are many other such representations possible for the section $f_{\alpha_0\ldots\alpha_q}$; it should be emphasized that a choice has been made here. If $U_{\alpha_0}\cap\ldots\cap U_{\alpha_q} \subset U_\beta$, then the same section $f_{\alpha_0\ldots\alpha_q}$ has a representation $f_{\alpha_0\ldots\alpha_q}(z_\beta)$ under the identification $\Gamma(U_{\alpha_0}\cap\ldots\cap U_{\alpha_q}, \mathcal{S}) \cong \Gamma(U_{\alpha_0}\cap\ldots\cap U_{\alpha_q}, \mathcal{O}^m)$ provided by the coordinate neighborhood $U_\beta \supset U_{\alpha_0}\cap\ldots\cap U_{\alpha_q}$; here $f_{\alpha_0\ldots\alpha_q}(z_\beta)$ is a complex analytic mapping from the set $z_\beta(U_{\alpha_0}\cap\ldots\cap U_{\alpha_q}) \subset \mathbf{C}$ into $\mathbf{C}^m$. These representations are of course related by

$$(3)\quad f_{\alpha_0\ldots\alpha_q}(z_\beta(p)) = \Phi_{\beta\alpha_q}(p)\, f_{\alpha_0\ldots\alpha_q}(z_\alpha(p)), \quad p \in U_{\alpha_0}\cap\ldots\cap U_{\alpha_q}\cap U_\beta .$$

The cocycle condition (2) can be rewritten as merely a condition on complex analytic functions, by using the identification $\Gamma(U_{\alpha_0}\cap\ldots\cap U_{\alpha_{q+1}}, \mathcal{S}) \cong \Gamma(U_{\alpha_0}\cap\ldots\cap U_{\alpha_{q+1}}, \mathcal{O}^m)$ provided by the coordinate neighborhood $U_{\alpha_{q+1}} \supset U_{\alpha_0}\cap\ldots\cap U_{\alpha_{q+1}}$; it has the form

$$(4)\quad \sum_{j=0}^{q+1} (-1)^j f_{\alpha_0\ldots\alpha_{j-1}\alpha_{j+1}\ldots\alpha_{q+1}}(z_{\alpha_{q+1}}) = 0 ,$$

$$z_{\alpha_{q+1}} \in z_{\alpha_{q+1}}(U_{\alpha_0}\cap\ldots\cap U_{\alpha_{q+1}}) .$$

Note that all except the last term in the sum (4) have a natural form, as an analytic function in terms of the coordinate corresponding to the last index. The last term can be rewritten

$$(5) \qquad f_{\alpha_0 \ldots \alpha_q}(z_{\alpha_{q+1}}) = \Phi_{\alpha_{q+1}\alpha_q} \cdot f_{\alpha_0 \ldots \alpha_q}(z_{\alpha_q}) .$$

Now, identifying the section $f_{\alpha_0 \ldots \alpha_q} \in \Gamma(U_{\alpha_0} \cap \ldots \cap U_{\alpha_q}, \mathcal{S})$ with the analytic function $\tilde{f}_{\alpha_0 \ldots \alpha_q} = f_{\alpha_0 \ldots \alpha_q}(z_q) \in \Gamma(U_{\alpha_0} \cap \ldots \cap U_{\alpha_q}, \mathcal{O}^m)$ the cocycle condition can be written

$$(6) \quad \sum_{j=0}^{q} (-1)^j \tilde{f}_{\alpha_0 \ldots \alpha_{j-1}\alpha_{j+1} \ldots \alpha_{q+1}}(p) + (-1)^{q+1} \Phi_{\alpha_{q+1}\alpha_q} \tilde{f}_{\alpha_0 \ldots \alpha_q}(p) = 0 .$$

Similarly, the coboundary condition

$$(7) \qquad (\delta f)_{\alpha_0 \ldots \alpha_{q+1}}(p) = \sum_{j=0}^{q+1} (-1)^j f_{\alpha_0 \ldots \alpha_{j-1}\alpha_{j+1} \ldots \alpha_{q+1}}(p)$$

in terms of the analytic functions representing the sections takes the form

$$(8) \quad (\delta \tilde{f})_{\alpha_0 \ldots \alpha_{q+1}}(p)$$
$$= \sum_{j=0}^{q} (-1)^j f_{\alpha_0 \ldots \alpha_{j-1}\alpha_{j+1} \ldots \alpha_{q+1}}(p) + (-1)^{q+1} \Phi_{\alpha_{q+1}\alpha_q} f_{\alpha_0 \ldots \alpha_q}(p) .$$

In particular, the group $H^1(\mathfrak{U}, \mathcal{Q}(\Phi))$ can be described, in terms of a representative cocycle $(\Phi_{\alpha\beta}) \in Z^1(\mathfrak{U}, \mathcal{GL}(m, \mathcal{Q}))$ as follows. The cocycles $Z^1(\mathfrak{U}, \mathcal{Q}(\Phi))$ are given by complex analytic mappings $f_{\alpha\beta}: U_\alpha \cap U_\beta \longrightarrow \mathbf{C}$ such that

$$(9) \quad f_{\beta\gamma}(p) - f_{\alpha\gamma}(p) + \Phi_{\gamma\beta}(p) \cdot f_{\alpha\beta}(p) = 0 \quad \text{for} \quad p \in U_\alpha \cap U_\beta \cap U_\gamma ,$$

or equivalently,

$$(9') \qquad f_{\alpha\gamma}(p) = \Phi_{\beta\gamma}(p)^{-1} \cdot f_{\alpha\beta}(p) + f_{\beta\gamma}(p) \quad \text{for} \quad p \in U_\alpha \cap U_\beta \cap U_\gamma \ .$$

The cocycle $f = (f_{\alpha\beta})$ is the coboundary of a zero-cochain $h = (h_\alpha)$ when the h_α are complex analytic mappings $h_\alpha \colon U_\alpha \to \mathbf{C}^m$ such that

$$(10) \qquad f_{\alpha\beta}(p) = h_\beta(p) - \Phi_{\alpha\beta}(p)^{-1} h_\alpha(p) \quad \text{for} \quad p \in U_\alpha \cap U_\beta \ .$$

Appendix 2. Some complications in describing classes of flat vector bundles.

Consider in particular the problem of describing all analytically trivial flat complex vector bundles of rank n over a compact Riemann surface M. According to Theorem 17, there is a one-to-one correpondence between this set of flat bundles and the set $\Lambda^*(\mathfrak{U}, I)$ of equivalence classes of complex analytic connections for the identity bundle I for a suitable open covering $\mathfrak{U}$ of the space M. For a general complex vector bundle Φ defined by a cocycle $(\Phi_{\alpha\beta}) \in Z^1(\mathfrak{U}, \mathfrak{GL}(n, \mathcal{O}))$, a connection is given by matrices λ_α of holomorphic differential forms of degree 1 in the various sets U_α, such that

$$(1) \qquad \lambda_\beta - \Phi_{\alpha\beta}^{-1}\lambda_\alpha\Phi_{\alpha\beta} = D\Phi_{\alpha\beta} = \Phi_{\alpha\beta}^{-1}d\Phi_{\alpha\beta} \quad \text{in} \quad U_\alpha \cap U_\beta \; ;$$

when Φ is the identity bundle, defined by the cocycle $\Phi_{\alpha\beta} = I$, this condition reduces to

$$(1') \qquad \lambda_\alpha = \lambda_\beta \quad \text{in} \quad U_\alpha \cap U_\beta \, ,$$

so that $\Lambda(\mathfrak{U}, \Phi) = \Gamma(M, (\mathcal{O}^{1,0})^{n\times n})$. An automorphism $T = (T_\alpha) \in \mathrm{Aut}\,(\Phi)$ is a collection of holomorphic non-singular matrices T_α in the various sets U_α, such that

$$(2) \qquad T_\alpha\Phi_{\alpha\beta} = \Phi_{\alpha\beta}T_\beta \quad \text{in} \quad U_\alpha \cap U_\beta \; ;$$

when $\Phi_{\alpha\beta} = I$, this condition reduces to

$$(2') \qquad T_\alpha = T_\beta \quad \text{in} \quad U_\alpha \cap U_\beta \, ,$$

so $T = T_\alpha$ is a global holomorphic matrix function on M. Since M is compact, T is necessarily constant; so that actually

$\text{Aut}(I) \cong (GL(n,\mathbf{C}))$. Two connections (λ_α) and (λ'_α) in $\Lambda(\mathfrak{U},\Phi)$ are equivalent precisely when there is an automorphism $T = (T_\alpha) \in \text{Aut}(\Phi)$ such that

(3) $$\lambda'_\alpha = T_\alpha\lambda_\alpha T_\alpha^{-1} - dT_\alpha T_\alpha^{-1} \quad \text{in } U_\alpha ;$$

and when $\Phi_{\alpha\beta} = I$, this condition reduces to

(3') $$\lambda'_\alpha = T\lambda_\alpha T^{-1} \quad \text{for } T \in GL(n,\mathbf{C}) .$$

Thus the analytically trivial flat vector bundles of rank n over M are in one-to-one correspondence with the equivalence classes (under conjugation by matrices $T \in GL(n,\mathbf{C})$) of $n \times n$ matrices of abelian differential forms on M. Given any matrix $\lambda \in \Gamma(M,(\mathcal{O}^{1,0})^{n\times n})$, select non-singular holomorphic functions F_α in the various sets U_α such that $\lambda|U_\alpha = DF_\alpha = F_\alpha^{-1}dF_\alpha$; the flat vector bundle associated to λ is that described by the cocycle $(\chi_{\alpha\beta}) \in Z^1(\mathfrak{U},GL(n,\mathbf{C}))$, where

(4) $$\chi_{\alpha\beta} = F_\alpha F_\beta^{-1} \quad \text{in } U_\alpha \cap U_\beta .$$

It is quite easy to describe the characteristic representation $\hat{\chi}$ of the bundle χ in a parallel manner. Let $f\colon \tilde{M} \longrightarrow M$ be the universal covering space of the Riemann surface M, and view $\pi_1(M,p_0)$ as the group of covering transformations, as in §7. Let $\tilde{\lambda} = f^*\lambda \in \Gamma(\tilde{M},(\mathcal{O}^{1,0})^{n\times n})$ be the matrix differential form on $\tilde{M}$ induced from λ by the covering mapping; the form $\tilde{\lambda}$ then satisfies $\tilde{\lambda}(\gamma\cdot\tilde{z}) = \tilde{\lambda}(\tilde{z})$ for all $\gamma \in \pi_1(M,p_0)$. Further, let F be a holomorphic non-singular matrix-valued differential form on $\tilde{M}$ such that

(5) $$dF = F\tilde{\lambda} ;$$

this function then satisfies

(6) $$F(\gamma \cdot \tilde{z}) = \hat{X}(\gamma) \cdot F(\tilde{z}) \quad \text{for all} \quad \gamma \in \pi_1(M, p_0) ,$$

where $\hat{X}: \pi_1(M,p_0) \longrightarrow GL(n,\mathbb{C})$ is a homomorphism representing the characteristic representation of the bundle X. (The function F can be viewed as arising from the functions on $\tilde{M}$ induced from the functions F_α under the covering mapping f, after the induced bundle $\tilde{X}$ on $\tilde{M}$ is reduced to the trivial bundle; compare with the discussion on page 145.) Note that the function F is uniquely determined only up to a constant factor $C \cdot F$ for any $C \in GL(n,\mathbb{C})$; this corresponds to the fact that the homomorphism $\hat{X}$ is determined only up to an inner automorphism in $GL(n,\mathbb{C})$. Recalling the discussion in §7 again, especially (22), the period class $\hat{A} \in H^1(\pi_1(M),\hat{I}) = \text{Hom}(\pi_1(M),\mathbb{C}^{n \times n})$ of the differential form $\lambda \in \Gamma(M,(\mathcal{O}^{1,0})^{n \times n})$ can be described in a similar manner as well. Selecting a holomorphic, matrix-valued function H on $\tilde{M}$ such that

(7) $$dH = \tilde{\lambda} ,$$

this function then satisfies

(8) $$H(\tilde{z}) - H(\gamma \cdot \tilde{z}) = \hat{A}_\gamma \quad \text{for all} \quad \gamma \in \pi_1(M,p_0) ,$$

where $(\hat{A}_\gamma) \in Z^1(\pi_1(M),\hat{I})$ is the period class of λ. Note that the function H is uniquely determined up to an additive constant, and hence the homomorphism $\hat{A}$ is uniquely determined by λ.

For flat bundles of rank $n = 1$, inner automorphisms are trivial, so both the characteristic representation $\hat{X}$ and the period class $\hat{a}$ are homomorphisms uniquely determined by the differential form $\lambda \in \Gamma(M, \mathcal{O}^{1,0})$. In this case, the relation between these two homomorphisms is particularly simple, which can be seen as follows. Comparing the differential equations (5) and (7), $\tilde{\lambda} = dh = f^{-1}df = d \log f$, so that the functions f and h are related by $f = ce^h$ for some constant $c \neq 0$; but then, applying equations (6) and (8), it follows immediately that

$$\hat{\chi}(\gamma) = e^{-\hat{a}_\gamma} \quad \text{for all} \quad \gamma \in \pi_1(M,p_o) . \tag{9}$$

Thus the analytically trivial flat line bundles are determined very directly in terms of the period classes of the abelian differentials; recalling Lemma 22, this can be rewritten in the more familiar form

$$\hat{\chi}(\gamma) = e^{\int_\gamma \lambda} \quad \text{for all} \quad \gamma \in \pi_1(M,p_o) , \tag{10}$$

where $\lambda \in \Gamma(M, \mathcal{O}^{1,0})$. (Recall also the discussion in §8 of last year's lectures.)

For flat bundles of rank $n > 1$, the characteristic representation $\hat{\chi}$ is determined only up to inner automorphisms in $GL(n,\mathbb{C})$, so one could not expect such a simple relationship as (9) to hold between the characteristic representation and the period class of a matrix $\lambda \in \Gamma(M,(\mathcal{O}^{1,0})^{n \times n})$ of abelian differentials. One might hope at least that the character of the representation $\hat{\chi}$ is determined directly by the period class $\hat{A}$, as a weaker form of (9); but unfortunately that is a vain hope as well. The characteristic representation $\hat{\chi}$ (which is actually an equivalence class of representations of the group $\pi_1(M)$) is of course uniquely determined by the period class $\hat{A}$, since both are uniquely determined by the matrix λ of abelian differentials. But the class $\hat{X}$ cannot be described directly as a function of the class $\hat{A}$ alone; the relation between $\hat{X}$ and $\hat{A}$ must involve the matrix λ, and hence the global structure of the compact Riemann surface.

To see that this is so, it suffices to examine the following simpler situation. Suppose $\lambda(z)$ is an $n \times n$ matrix of holomorphic differential forms of degree 1 in an open neighborhood U of the real axis in the complex plane, such that $\lambda(z+1) = \lambda(z)$; and let $F(z)$ be a non-singular holomorphic matrix function in U such that $dF = F\lambda$, and H be a holomorphic matrix function in U such that $dH = \lambda$. These two functions then satisfy $F(z+1) = X \cdot F(z)$ for some matrix $X \in GL(n,\mathbb{C})$, and $H(z+1) = A + H(z)$ for some matrix $A \in \mathbb{C}^{n\times n}$; F is uniquely determined up to a constant factor $C \cdot F$ for $C \in GL(n,\mathbb{C})$ and H is uniquely determined up to a constant term $H+B$, so that the matrix X is determined up to an inner automorphism in $GL(n,\mathbb{C})$ while A is uniquely determined by the differential form λ. (In the compact Riemann surface case as above, the universal covering space $\tilde{M}$ can be taken to be the upper half-plane, and the transformation γ can be taken in the form $\gamma \cdot \tilde{z} = c\tilde{z}$ for some real constant c; the exponential mapping reduces this to the special case envisaged here.) The problem is to find the extent to which the conjugacy class of the matrix X can be determined as a function of the matrix A alone.

Select a constant matrix S such that $e^S = X$, and

introduce holomorphic functions $G(z)$ and $K(z)$ in U such that

$$F(z) = e^{S\cdot z}G(z) \quad \text{and} \quad H(z) = A\cdot z + K(z) \; ; \tag{11}$$

it is easy to see that these functions $G(z)$ and $K(z)$ then satisfy

$$G(z+1) = G(z) \quad \text{and} \quad K(z+1) = K(z) \; . \tag{12}$$

Such functions admit a Fourier expansion in a neighborhood of the real line, of the form

$$K(z) = \sum_{n=-\infty}^{+\infty} K_n e^{2\pi i n z} \tag{13}$$

for suitable matrices $K_n \in \mathbf{C}^{n\times n}$. Recalling that $\lambda(z) =$
$= F(z)^{-1}\, dF(z) = G(z)^{-1}\, e^{-S\cdot z}[e^{S\cdot z}\, dG(z) + e^{S\cdot z}\, S\cdot dz G(z)] =$
$= G(z)^{-1}\, dG(z) + G(z)^{-1}\, SG(z)dz$ and $\lambda(z) = dH(z) = A\cdot dz + dK(z)$,
it further follows that

$$A + \frac{d}{dz}\, K(z) = G(z)^{-1}\, \frac{d}{dz}\, G(z) + G(z)^{-1}\, SG(z) \; . \tag{14}$$

Again both sides of (14) are invariant under the translation $z \longrightarrow z+1$, so admit Fourier expansions of the form (13). For the left-hand side in particular, the expansion is just

$$A + \sum_{n=-\infty}^{+\infty} 2\pi i n\, K_n e^{2\pi i n z} \; ,$$

so that A is the constant term; and thus

$$A = \text{const}[G(z)^{-1}\, \frac{d}{dz}\, G(z) + G(z)^{-1}\, SG(z)] \; , \tag{15}$$

where $\text{const}[\cdot]$ denotes the constant term of the Fourier expansion of the expression in brackets. This formula expresses A in terms of the matrix S, hence of course in terms of X; but the expression involves in addition the function $G(z)$, which can be an arbitrary nowhere-singular analytic matrix function invariant under the translation $z \longrightarrow z+1$. The discussion is now reduced to seeing the extent to which the choice of the function $G(z)$ affects the relationship between A and S in (15).

Note first as a consequence of (15), that, letting tr denote the trace of a matrix

$$\text{tr } A = \text{tr } S + \text{const } [\text{tr } G(z)^{-1} \frac{d}{dz} G(z)] \; . \tag{16}$$

However $\text{tr } G(z)^{-1} \frac{d}{dz} G(z) = \frac{d}{dz} \log \det G(z)$; and since $\log \det G(z)$ is invariant under the translation $z \longrightarrow z+1$, its derivative has a Fourier expansion in which no constant term appears. Therefore (16) reduces to

$$\text{tr } A = \text{tr } S \; ; \tag{17}$$

or in other words, $\det X = e^{\text{tr } S} = e^{\text{tr } A}$ is a relation between A and X which does not depend on the choice of function $G(z)$. To see that this is in general the only such relation, consider in the 2×2 matrix case the function

$$G(z) = Me^{-2\pi iz} + I + Me^{2\pi iz} \quad \text{where} \quad M = \begin{pmatrix} -a & -a^2 \\ 1 & a \end{pmatrix} \tag{18}$$

for some constant $a \in \mathbf{C}$; the function $G(z)$ is invariant under the translation $z \longrightarrow z+1$, and indeed, (18) is just its Fourier expansion. This can be rewritten

$$G(z) = \begin{pmatrix} 1 - 2a \cos 2\pi z & -2a^2 \cos 2\pi z \\ 2 \cos 2\pi z & 1 + 2a \cos 2\pi z \end{pmatrix} ,$$

from which it is apparent that $G(z) \equiv 1$. For any matrix

$G = \begin{pmatrix} \alpha & \beta \\ \gamma & \delta \end{pmatrix}$ with $\det G = 1$ it is obvious that $G^{-1} = \begin{pmatrix} \delta & -\beta \\ -\gamma & \alpha \end{pmatrix}$; and therefore

(19) $$G(z)^{-1} = M'e^{-2\pi iz} + I + M'e^{2\pi iz} \quad \text{where} \quad M' = \begin{pmatrix} a & a^2 \\ -1 & -a \end{pmatrix} .$$

Now it is clear that

$$\text{const } [G(z)^{-1} \frac{d}{dz} G(z)]$$

$$= \text{const } [(M'e^{-2\pi iz} + I + M'e^{2\pi iz})(-2\pi i\, Me^{-2\pi iz} + 2\pi i\, Me^{2\pi iz})]$$

$$= M' \cdot 2\pi i M + M' \cdot (-2\pi i M) = 0 ;$$

and that

$$\text{const } [G(z)^{-1} SG(z)]$$

$$= \text{const } [(M'e^{-2\pi iz} + I + M'e^{2\pi iz})S(Me^{-2\pi iz} + I + Me^{2\pi iz})]$$

$$= S + 2M'SM .$$

Therefore (15) reduces to the equality

(20) $$A = S + 2M'SM ,$$

which explicitly involves the matrix M depending on an arbitrary parameter a . In particular, taking $S = \begin{pmatrix} s_1 & 0 \\ 0 & s_2 \end{pmatrix}$ for example, (20) becomes

(21) $$A = \begin{pmatrix} s_1 - 2a^2(s_1-s_2) & -2a^3(s_1-s_2) \\ 2a(s_1-s_2) & s_2 + 2a^2(s_1-s_2) \end{pmatrix} .$$

Thus, $\operatorname{tr} A = s_1+s_2 = \operatorname{tr} S$, and

(22) $$\det A = \det S + 2a^2(s_1-s_2)^2 ;$$

so if $s_1 \neq s_2$, $\det A$ can be made arbitrary by suitable choice of the parameter a , showing that the eigenvalues of A can be arbitrary subject only to the restriction that $\operatorname{tr} A = \operatorname{tr} S$.

This observation shows that one cannot expect the description of analytically trivial flat vector bundles to be as straightforward as the description of the analytically trivial flat line bundles, as given in §8 of last year's lectures for instance; and this may explain some of the complications in the present discussion.

Lightning Source UK Ltd.
Milton Keynes UK
UKHW010148250921
391135UK00001B/41

9 780691 079981